STUDENT RESEARCH PROJECTS
IN CALCULUS

Marcus S. Cohen
Edward D. Gaughan
Arthur Knoebel
Douglas S. Kurtz
David J. Pengelley

SPECTRUM SERIES

The Spectrum Series of the Mathematical Association of America was so named to reflect its purpose: to publish a broad range of books including biographies, accessible expositions of old or new mathematical ideas, reprints and revisions of excellent out-of-print books, popular works, and other monographs of high interest that will appeal to a broad range of readers, including students and teachers of mathematics, mathematical amateurs, and researchers.

Mathematical Association of America
1529 Eighteenth Street, NW
Washington, DC 20036
800-331-1MAA   FAX 202-265-2384

Spectrum Series

STUDENT RESEARCH PROJECTS IN CALCULUS

MARCUS S. COHEN
EDWARD D. GAUGHAN
ARTHUR KNOEBEL
DOUGLAS S. KURTZ
DAVID J. PENGELLEY

MAA
SPECTRUM

Published and Distributed by
THE MATHEMATICAL ASSOCIATION OF AMERICA

Preface

You are reading this preface because you are curious about student research projects in calculus. Thus you are a member of our intended audience: that growing group of people sincerely interested in improving their students' knowledge of calculus. As you read on, you will discover new ways to challenge your students and stimulate their interest in calculus, as well as yours.

The introduction gives a brief description and the guiding philosophy of student research projects in calculus. Chapter 1 is devoted to the history and evolution of our program. In Chapter 2, we discuss our evaluation of the effects of the program over the past several years. Chapter 3 is devoted to the logistics of using projects in the classroom. In this chapter we summarize our experiences of the past years and point out those strategies that work and those that do not work. After reading these chapters, we hope you will get into the spirit of the movement and, not only will you want to try projects in your calculus classes, you will want to try your hand at creating some projects of your own. Chapter 4 will offer some advice about how to write projects. Read on, anticipate, and let your enthusiasm grow.

The second part of this book is a compendium of projects, most of which have been class tested. You have our permission to reproduce any of these projects for classroom use. Each project includes a commentary and some additional information concerning that project. The "Guide to the Projects" section gives you information about using the project section of the book and the related index.

ACKNOWLEDGEMENTS

One worries a bit when starting to thank those who contributed to a book. Some inclusions are obvious, but there is always a danger that someone may be overlooked. We apologize in advance to anyone we may omit by accident.

First of all, we must thank Louise Raphael of Howard University, who was our program director at the National Science Foundation in the spring of 1988. Without her untiring guidance through proposal and budget revisions, our program and this book would never have seen the light of day. We owe her much. Our thanks also go to the National Science Foundation for the funding of our Calculus Curriculum Grant.

In planning our proposal to the National Science Foundation, we decided to experiment with projects at some neighboring schools. The results of our experiment were skewed by the fact that we chose faculty who are very interested in good teaching and are outstanding teachers. Initially, we referred to them as "teaching consultants," thought of them as "trainees," and, after they had a semester of experience, we came to consider them as equal partners in our endeavors. Their experiences and contributions were very valuable. We wish to thank: David Arterburn of the New Mexico Institute of Mining and Technology, Adrienne Dare of Western New Mexico University, Carl Hall of the University of Texas at El Paso, and Richard Metzler of the University of New Mexico.

Over the past three years, a number of graduate students have been our willing assistants and collaborators. They have contributed much to our program and we hope that we have offered them some unique teaching experience. They suffered through crowded labs, patiently counseled impatient students, and participated in our post-project discussions. We wish to thank: Randel Combs, Larry Hughes, Gamal Hussien, David Ruch, Christopher Stuart, and Cynthia Woodburn.

The evaluation chapter was written by Marsha Conley and Carol Stuessy. Marsha was a member of our faculty and was assigned part-time to our project to assist in the gathering of data. Carol is on the faculty at Texas A & M University and has been our faithful evaluator from the beginning of our project. Further evaluation was provided by Gabriel Lampert, a statistical consultant at New Mexico State University, during the summer of 1991. We extend them our sincere thanks for all their efforts.

Our colleague Jerry Lodder is the author of two of our projects, *The instantaneous value of the bending ratio* (#34)[1] and *Estimating bounds for the bending ratio* (#35). Two colleagues in other departments, William Boecklin in biology and Barry Kurtz in computer science, helped create a half dozen of the projects, some of them using computers. We thank them for their contributions and enthusiasm.

We also wish to thank Spectrum editor James Daniel and the reviewers for their excellent and thoughtful suggestions and advice. Our thanks also go to Donald Albers, Chair of the Committee on Publications, for his interest in the manuscript and his guidance that brought this book to publication.

Finally, we must thank our Department Head, Carol Walker. Without her support, advice, and interest, our task would have been almost impossible.

---

[1] The project number refers to the "List of Projects."

Contents

INTRODUCTION

This compendium of calculus projects represents the result of three years of creating, testing, and fine-tuning an approach to teaching calculus designed to:

1. Develop independent analytical thinking by combining intuition, generalization and logical argument;

2. Build the skills to mathematically model a situation by defining variables, establishing relations between them, and working deductively towards an end result;

3. Cultivate an appreciation of mathematics as an evolving entity created and discovered by human enterprise.

A project is a multistep take-home assignment that typically takes two weeks of concerted effort to complete. Projects may be assigned to individuals working alone or to groups of students. There are places in each project where we expect even the best students to get stuck. Sometimes students get unstuck by a sudden insight of their own; more often, they seek help from their instructors or in a project lab, designed especially for this purpose. This help is given in the form of questions, hints and/or supplementary exercises and reading. While many of our projects are woven around a "real-world" story-line, we do not attempt to give a completely rigorous account of the relevant background material from physics, biology, or engineering. Rather, each project is designed to help a student explore some part of calculus and discover its utility. Many projects guide students to construct a formal proof, but often only after exploring one or more examples and giving them an intuitive feeling for the context of the result.

The structure of these project problems and the infrastructure supporting their use in the classroom was motivated by two pedagogical premises:

1. Students learn and retain mathematics better if they explore it on their own, passing from specific examples to general principles;

2. Provided that a significant portion of their grade depends on doing so, students will be motivated enough to follow this guided exploration, filling in the missing steps for themselves along the way.

Each project may be used as a window into the highly interconnected structure of the calculus; if several projects are used in a semester, they will hopefully provide solid islands of understanding from which students may work to connect the structure in their own minds. This understanding is acquired through guided self-discovery of the proper tools and techniques to solve each project problem. Certainly such discovery learning is better than treating a student's mind as a tabula rasa onto which the instructor writes a completed and logically consistent theory, followed by exams to reinforce correct transcription and punish mistakes. People learn by doing, by exploring and being guided when perplexed, and by caring about obtaining a positive result (even if only a good grade) from their explorations. To what extent guided self-discovery must be supplemented by worked examples and drill remains to be seen.

This book of projects is intended to encourage you to use such discovery-learning in your calculus sequence. Instructors must provide support material and catalyze the interconnection of the concepts learned in projects with material in the text and in lectures. We have used two to five projects per semester; some of us have distributed notes and assigned supplementary problems to complement and reinforce the projects, while others stuck fairly closely to the departmental syllabus. We all noticed that students were more interested in lectures, and in examples and problems worked in the classroom, when they thought that these might give some insight into their projects. We often found it effective to time lecture material so questions are lit in students' minds by their projects *before* the lecture, and that the lecture then helps answer those questions.

We do not suggest any particular syllabus of lectures and discussions here. We have tried to provide a broad enough range of projects to use in any syllabus.

Yes, project-based calculus does require more time of both instructors and students. Extra office hours must often be scheduled to help students with their projects. However, it is fun and energizing to give appropriate hints to students who come in after pondering long and hard the statement of a problem, trying perhaps several approaches on their own and getting stuck, as opposed to responding to the usual panic-stricken inquiry, "Will this be on the exam?" Grading projects is also much less of a chore than grading exams, since you are seeing students' best work, instead of their worst. Since algebraic errors have usually been checked and corrected by the time a project has been handed in, you are free to grade for conceptual content and logical flow rather than purely on technical correctness.

Students seem willing to invest incredible amounts of time and energy on their projects (often at the expense of other courses!), especially students who characterize themselves as "not testing well." They see an opportunity to be graded on their achievements rather than on timed tests, and they respond in kind. Many projects contain "extra credit" parts designed to challenge good students to excel.

We encourage you, as instructors, to view this book as the vehicle of an evolving participatory endeavor rather than as a recipe set in stone. To make it easier to access the main topics you may want a project to cover, we have provided an index to the projects; this index references prerequisite ideas that feed into and themes that branch out of each project. If you have had enough background material for now and want to proceed directly to the selection of a project for your class, you may turn to the "Guide to the Projects." However, before you actually assign your first project, please read the section on "Logistics: Ideas for Using Projects Successfully"—it will at least help you avoid the pitfalls we fell into! A chronicle of our own experiments and experiences using projects in our calculus classes appears in the section entitled "History."

A digest of our experience that may prove useful for writing your own projects appears in the section entitled "Creating Projects." Finally, the "Evaluation" section gives summary qualitative and quantitative evaluations of our success to date with the project approach.

Have fun!

Part I

Chapter 1

HISTORY

In 1986, 700 students were enrolled in the three semester mainstream calculus sequence at New Mexico State University. Each lecture class had approximately 40 students, met 150 minutes per week, and grading was generally based on hour examinations, homework and quizzes, and a comprehensive final examination. In 1991, students took the same courses, but with one major difference. In 80% of the classes, a significant part of each student's grade was based on performance on student research projects. We hope that by telling you how we got from 1986 to 1991, you will learn something about how to use student research projects yourself

In the summer of 1987, Marcus Cohen and David Pengelley met to discuss the upcoming semester. They were both to teach the third semester calculus course for science and engineering students. The general downward trend of students' performance over previous years had made this less enjoyable and they wondered how to motivate students to care about understanding the concepts and mastering the skills of the calculus. They decided to try to recreate for students the experience of the research mathematician, who is motivated by an interesting problem. They decided to assign challenging multistep problems with some difficult points, and see if students would make these problems their own, learn the relevant background material, and seek help when stuck. The idea for student research projects was born.

They were very excited and decided to try to make up some truly difficult and challenging problems. They set up shop in the midst of calculus texts strewn on the floor of Marcus Cohen's living room and began working. In a short amount of time they had developed ideas for several projects, and they were hooked.

That first semester was exciting for them, and a lot of work! (Fortunately, we know *much* more now than they did back then.) They had decided to make three project assignments, and write six different projects for each one. But writing six projects of equivalent difficulty takes time. Still, by the end of the semester, they were convinced that assigning student research projects was the thing to do.

The traditional means of evaluating students' performance in mathematics classes has been to use examinations, and after watching students perform poorly on examinations for years, we tend to think less of our students' abilities. What Marcus Cohen and David Pengelley learned was that many of their students did not test well but could solve difficult problems. The projects were also having a profound effect on the students. On surveys, students often made statements like "It really made me feel good to be able to solve a problem by myself. It gave me confidence" and "I did this and I'm proud of it."

The results were so exciting that they wanted to tell the world, at least locally. The two of them held an open forum to tell the rest of the department what they had learned. Some people were interested, but the response was less than overwhelming. Still, like any good virus, the use of projects began to spread. During the spring semester of 1988, Doug Kurtz and graduate student Dave Ruch began to use projects as well. At about this time, the National Science Foundation put out its first request for proposals to address the learning of calculus. Marcus Cohen, Doug Kurtz, and David Pengelley thought that the use of student research projects might be one good approach to the problem and began to think about submitting a proposal. They wanted a larger group of participants for such an undertaking and induced Ed Gaughan and Arthur Knoebel to join the team.

Much of the spring semester was spent writing the proposal and getting more people involved. We included faculty members from some nearby universities: Dave Arterburn of New Mexico Institute of Mining and Technology, Adrienne Dare of Western New Mexico University, and Carl Hall of the University of Texas at El Paso. A year later we were joined by Dick Metzler of the University of New Mexico. To test the efficacy of our program, we convinced Carol Stuessy of Texas A&M University to be our educational consultant and the department assigned Marsha Conley as our statistical consultant. All of these people have been instrumental in making our program successful. The NSF was not going to leave this to chance and requested we create an advisory board. Paul Bamberg of Harvard University, Ronald Douglas of the State University of New York at Stony Brook, Murray Klamkin of the University of Alberta, and Alan Schoenfeld of the University of California at Berkeley gave us valuable insight and useful support.

The fall of 1988 saw the official beginning of a projects program at NMSU, with projects used in seven calculus classes taught by four faculty members and three teaching assistants. Ed Gaughan and Arthur Knoebel were using projects for the first time, and by the end of the semester, the latter had uttered the now famous battle cry, "It is amazing how one's opinion of students goes up after using projects." It is hard to explain how exhilarating it is to watch students succeed at something beyond what we (and they) perceive as their abilities. It has an equally strong effect on the students, "The most enjoyable part is knowing that you are capable of solving such a problem."

We adjusted our sights after a year of using projects and decided to make two project assignments per semester, handing out three projects per assignment. It became obvious to us that, if we wanted to create a program that could be used at other schools, we would have to devise a method of making the demand on instructors' time manageable. This led to our most useful and frustrating invention, a student help lab.

Serendipity has played a large role in the development of our program. If Marcus Cohen and David Pengelley had not lived across the street from each other, they might never have had their conversation in August, 1987. And, if Doug Kurtz had not been in his office when David Pengelley walked past in May, 1988, we might never have come up with the idea for our student help lab staffed primarily by teaching assistants. The lab was open ten to fifteen hours per week during the project assignments for students to ask questions about their projects. We will discuss this lab later.

During November we held our first workshop for program participants. In addition to the people who would be teaching with projects, the workshop was attended by Carol Stuessy and Louise Raphael, the first program director at NSF for the calculus initiative. The main goal of this workshop was to begin to prepare the people from the other universities to use projects during the spring semester. A second workshop was held in January of 1989. The previous one had served to whet the appetites of our cohorts from other universities, and this one was to teach them the day-to-day activities. In response to their comments about the previous workshop, we included role playing sessions about handing out projects, conducting office visits and interviews, and grading. The first workshop addressed theory; the second one practice. Then we sent them off to learn through "trial by fire." Miraculously, it worked! (We should really be honest about why it worked. If you choose good, hard working people and keep out of their way, good things happen. That was one of our main strategies.)

There was one notable event of the first semester that profoundly affected our program. For the second round of projects, we had assigned the project *Houdini's escape* (#40)[2]. The main goal of this project was to get students to understand the second fundamental theorem of calculus. We had found that students reacted well to story lines, so we built a story around the mathematics in the project. We did not view this problem as applied; we viewed it more as cute. However, on surveys at the end of the course, many students spoke about how much they liked this *applied* problem. This had a major effect on the projects we wrote the next year. The students seemed to view problems they could visualize as applied problems, and they responded more favorably to them. We began to write more story lines into our projects. This helped to attract students' attention and it also made many of the projects harder, since the problems were buried in text. A good example of this is *Calculus in the courtroom* (#32); there does not appear to be much mathematical guidance in the project itself. And the students seemed to be comfortable with that.

---

[2] The number of a cited project, to be found in Part II, is given here in parentheses.

The students' reaction to the story lines was really amazing. Their reports often extended the story, sometimes including dialogue. One student consistently handed in artwork related to the stories of her projects. We think this says quite a bit about the students' feelings of pride and confidence in their work.

The first semester went relatively smoothly, so we were optimistic as the next semester began. We were going to offer nine projects classes and have about 100 extra students, for a total of 340, but we were ready. Or so we thought!

To *simplify* the work for graduate students running the projects lab, we arranged for the projects for the three different calculus courses to be offered at the same time. Three different projects were given out in each course, so each teaching assistant was required to provide assistance for *nine* projects simultaneously. (Not a simple task; not all of the professors knew how to solve all of them!) And, after the previous semester, students had learned that the best way to solve a project was to get lots of help in the lab. Many more students used the lab the second semester, and we had not scheduled the lab to handle this increase. This was truly the "lab from hell." Most of the time there were two or three people working in the lab (one faculty and the rest teaching assistants), and thirty to forty students waiting to ask questions. Not everyone working in the lab could answer questions about all the projects, which was frustrating for students and embarrassing for instructors. That was the first week, and the second week was worse! We have seen our students learn about projects over the last two years, so that now they know that they really do have to start as soon as they get their project. But a year ago, that was not so clear and many students put off working on the assignment for over a week.

That second semester, the situation was something like this. The project was due in three days and some student had just gotten around to working on it. She went to the lab to get help and had to wait for about an hour to have someone respond to her question by asking another question. (See the section "Logistics: Ideas for Using Projects Successfully" for information on the operation of our lab.) After two days of such *help*, some emotions were frazzled. In fact, it was really worse than it sounds. Few students will be abusive with their own instructor, but since students were often talking to some anonymous teaching assistant, many were rude. The situation was not pleasant for students, teaching assistants, or us. In part because of this, we made two changes that have provided tremendous benefits.

The first change we made was for mostly pedagogical reasons but had a great effect on simplifying the lab. David Pengelley, who was spending the year on sabbatical leave but still exerting his influence, would call about once a week and tell us we should have students work in groups to solve projects.

We spent a lot of time talking about this and resisted the idea, mainly because we did not have a clear idea of how we could evaluate the performance of individuals if they worked in groups. Still, we liked the idea for its educational and intellectual values and decided to try it and see what happened. We settled on three people to a group (although some worked in groups of two, some alone). Some quick arithmetic shows that when you divide three hundred students into groups of three, it is like working with only one hundred students, a much more manageable number. Also, we felt the need to assign only a single project per course for a group project. The effect on the lab was dramatic. People working in the lab only had to know how to solve three problems and only had to deal with one hundred groups. We have been very pleased with group projects and highly recommend their use. We discuss later the second change, which was to offer individual labs for each course.

Students either loved the group projects or hated them! Working in a group was an alien activity for most students and they often found scheduling time and making arrangements difficult. Still, many of them saw that the benefits easily exceeded the disadvantages. One student's quote sums it all up:

> The group projects were even more important because they encouraged the development of skills that are essential, such as the skills of communication. Actually talking about calculus helped me immensely in grasping concepts and understanding how real mathematicians work. I also gained much from seeing the different ways in which the other group members approached calculus. For the first time since elementary school, I was actually working in a group! Time went to developing skills that were important to calculus such as seeing other people's reasons, explanations, and approaches to mathematical ideas, learning to verbalize mathematical ideas, learning to communicate and work effectively in a group to achieve some mathematical goal.

While all this activity was going on at New Mexico State University, three people were using projects at distinctly different universities. Carl Hall was using projects at a school where over half the students come from minority groups and Adrienne Dare was working at a small university for teachers. But the most notable endeavor was by Dave Arterburn. His university attracts academically solid students, so he was willing to really experiment. He decided to assign each of his students *five* projects. He had a very small class, which allowed him to do this, but he was working on projects the entire semester!

One important point should be made here. While there was a cohesive group of seven to ten people working on projects at NMSU, the people working at other places were working alone. Carl Hall and Dave Arterburn did have some graduate student help, but no other collaborating faculty members. Adrienne Dare was running the whole show herself off in the mountains near the Gila Wilderness. Their experiences along with ours have shown us the benefits to be gained when several people work together on projects classes.

During the summer, we held our third workshop to get reactions from these people and plan for the second year. Looking back, we were all very pleased with the previous academic year. But Dave Arterburn was burned out. He was not even sure he wanted to use projects again. We all admired what he had done, being sure in our own minds that we would never do anything as foolish as he. We spent two days talking and planning, and trying to get Dave Arterburn to use projects again but on a more reasonable scale. By the second afternoon, we had planned the upcoming semester. Because we perceived that students in the first semester had a harder time with projects than students in later courses, the NMSU faculty would use two one-week individual projects and a two-week group project in our first semester of calculus, and a two-week individual project and a two-week group project in our second and third semesters. Carl Hall and Adrienne Dare would follow the latter scheme—and Dave Arterburn would use *five projects again*! (Some people are just incorrigible.) These three people had been using projects for only one semester, but already they had changed from trainees into our peers. This workshop was also attended by Dick Metzler, who was going to start using projects in the fall. Not only would his students solve student research projects, he had already organized his calculus class around a graphing, programmable calculator. Dick Metzler was the first to write and use projects regularly with calculators. (Sporadically, Arthur Knoebel was directing the writing of projects for both computers and calculators.)

It was also at about this time that we began to seriously edit the projects we had written and discuss what was a good project and what was a bad one, what worked and what did not. We found that it was easier to write projects in pairs and we suggest that people work that way. This simplifies the writing and solving and helps to uncover difficulties with the assignment. Each project that was used in the classroom (about 80% of the ones in this book) was edited after the project was assigned so we could incorporate students' reaction and understanding in improving the statement of the problem.

The use of projects has led to growth among the participants in the program. We have been forced to come to grips with educational issues. One fundamental issue involves the evaluation of students. We began with a real concern about cheating and individual performance. We have evolved to be less concerned about this. In fact, we now look for ways to encourage students to interact, even at the expense of evaluation. We had originally assigned multiple projects (first six at a time, then three) to discourage cheating. By the fall of 1989, we were willing to assign a single project to a course with as many as five sections.

In the fall of 1989, we implemented the second major change to our lab. This was to offer a separate lab for each course, which has many advantages. Projects can be assigned at optimal times for a particular course without regard to other courses, the number of students a lab serves is significantly smaller, and the people working in the lab only have to know material for one course, usually a course they are teaching. We also decided to close the lab several days before the projects are due. Thus, near the end of an assignment, the only person students can see for help is their own instructor, which helps to alleviate some of the emotional problems we had observed. All in all, these changes have made the lab a much more pleasant place.

We made other minor adjustments with our lab, some of which worked and others that are still questionable. One of the main complaints from students was the amount of time they had to wait in the lab to get help. We tried offering a "drop-in" lab and a "by appointment only" lab. Students could schedule a ten minute appointment at a specified time and not have to wait around for help. Surprisingly, students took very little advantage of this latter lab. During the spring semester of 1990, we began to use undergraduate students as assistants in the labs. We feel this can be successful, and look forward to making more use of undergraduates in the future, but we have learned that it is crucial that they be carefully trained and prepared for each lab. We are anxious to try using students who learned calculus in projects classes as lab assistants, since they have already been exposed to the process and will know the benefits of the program.

Although we had decided that it was sufficient to assign a single project at a time, this past year we began to write sets of related projects. An example is the trio *An increasing function and the most important limit in calculus* (#14), *A bounded function and the most important limit in calculus* (#15), and *The most important limit in calculus exists!*(#16), centered around the definition of the number e. After these projects had been graded and returned, some class time was spent discussing how the result of the projects could be combined to create e. The project *Preparing for the 1990 United States census* (#19) was assigned along with a similar one using New Mexico census data. In this case, the data for the United States led to a fairly good estimate for the population while the data for New Mexico did not.

It was easy for us to see the general trend in the level of the projects we wrote: they continued to get harder. If you do not expect much from students, you do not ask for much and that is what you get. As we learned that our students could do more, we asked for more. Each semester our projects have gotten more challenging!

Towards the end of the Fall semester, we began to look to the future. Our grant from NSF was coming to an end and there were questions for us to ponder. One involved the continuation of a projects program at New Mexico State University. Another concerned whether we were interested in pushing such a program in other directions.

An outgrowth of discussions about the future led us to consider the advantages of a departmental structure to oversee the use of projects. (For lack of a better word, we referred to this as the *institutionalization* of our program, but since that conjures up images we would rather not have associated with what we are trying to do, we will not use it further.) By the end of the 1989–1990 academic year, there were several changes in our department, such as the creation of a position of "projects coordinator" to coordinate the program across the courses, including training, scheduling, staffing the labs, and assisting the writing of projects.

We have begun to use projects in a differential equations course. In this course, designed by Marcus Cohen, projects are used as one of the central pedagogical instruments, introducing students to some of the main subject matter. A further goal is to unify the material of vector calculus with differential equations.

Another direction involves the use of projects in high schools. Long-term writing assignments are alien to students in mathematics classes. We have seen how students' skills at solving and writing solutions of projects improve with experience, and also how projects attract and allow many students to continue pursuing careers heavily dependent on mathematics. We think that these benefits would all be seen in a high school setting and, in fact, the use of projects may be more beneficial there. We are training high school teachers to allow them to use student research projects of an appropriate level in their classes.

The summer of 1990 saw the fifth and last workshop of our program. Some of the comments that came out of this workshop are illustrative of all our experiences. Dick Metzler was very pleased with the atmosphere in his class and felt that students will have good memories of their mathematics class. That is not a misprint; we have several stories we could tell. There is one about a student who solved his project while driving 225 miles to a heavy metal concert; and another about a group staying up all night working on their project at a local restaurant. Such stories do not sound surprising after you have had a student say to you, "Some nights, I couldn't sleep because I was thinking about how to approach a certain aspect of my project." One student gained insight into solving *Jacobi's Pizzeria* (#100) while—of course—eating a pizza. Dave Arterburn mentioned that students taking a projects class for the second or third time wanted *harder* projects! Jerry Lodder, a faculty member at NMSU, told how he used the idea of a "project" to introduce material during lectures; instead of proving "theorems," he would consider a "project" to find a formula for the derivative of a quotient. We invited a couple of students who had done projects in three classes to the workshop to get their (admittedly biased) opinions. What we found surprising was their ability to recall the subject matter of projects they had worked on over a year earlier. Imagine a student remembering anything positive about a test after a year!

We have learned a lot over the past several years and think we have an idea that is useful for teaching calculus, and in fact any mathematics course. We are offering workshops around the country, to help others learn to use projects and avoid some of the pitfalls of the process. Already, people not associated with our program have tried to use projects at their schools and we have learned from their experiences. There are two things that we did that had a large effect on making our program work. We made each project count as much as a test, generally 15% to 20% of a student's grade. We also gave students a definite time limit. These factors made it necessary for students to take the assignments very seriously. Others have tried to use projects as optional assignments and, unfortunately but not surprisingly, had mixed success.

One of the most encouraging things about using projects is that students have begun to be more interested in their mathematics courses. We have gotten to know our students better, personally and intellectually. Our students have gotten a glimpse of what attracts mathematicians to mathematics and many have found it enjoyable. Over the past two years, we have recruited more mathematics majors from our calculus classes than in the past, including some of the better students. This is an unexpected benefit.

Chapter **2**
EVALUATION*

EVALUATION CONTEXT

We have developed and evaluated an optional version of mainstream calculus at New Mexico State University. The form of the program incorporates several features that we consider to be typical of how such curriculum change would be implemented in most colleges and universities, but which restrict the scope of statistical interpretation of the results. Specifically, students were allowed to exercise self-selection in choice of enrollment in project or nonproject sections and there was no attempt at replication of section allocation to isolate the effect of individual instructors. Consequently, we are cautious in interpreting the exact role of the project program when comparing project and nonproject sections, and also cautious in identifying the mechanisms underlying the observed patterns. Inferential statistics were calculated for the various comparisons of interest, but given the limitations of the project design, we prefer an emphasis on descriptive comparisons. It is important to note that the results can also be viewed at many scales. The scale that is most commonly examined by an individual faculty member is that of individual sections in each semester. It is on this scale that the individual instructor forms a personal evaluation of student composition and the results of a particular teaching methodology. In this study, we repeatedly observed great variation in grade distributions between sections that were using projects, as well as between sections that were not using projects. It is clear that the effect of individual instructors and section composition still account for a great deal of the variation in student performance at the scale of semester periods. In the following summary, we

---

* An update based on an evaluation during the summer of 1991 appears at the end of this chapter.

emphasize large-scale patterns spanning the two-year course of the study. Given the large amount of variability between sections within a semester, and among semesters, this appears important if the results are to be of value when considering curriculum revision.

We feel that it is important to point out that "formal evaluation" as defined by course grades is an inadequate form of evaluation for the objectives of this program because course grades, even in project courses, were still based primarily on performance on traditional exams. By definition, the project experience is designed to teach features of mathematics that are antithetical to the exam setting. Thus the positive effects of the program that we are seeking may require very different measures of success. However, we are also aware that most faculty and administrators (and students!) will be curious about the program's track record as measured by traditional forms of evaluation, i.e., course grades. Therefore, we have conducted an extensive examination of this aspect. In analyses of student outcomes, we attempted to address the kinds of questions that we felt to be of most obvious concern when considering implementation of such a program. The following list presents those questions for which we give some answers:

1. How did pass rates observed in project and nonproject sections compare to pass rates during recent comparable semesters prior to the introduction of the project option?

2. Did the use of a project option as implemented in this study produce any long-term pattern of differences between project and nonproject classes in course grade distributions and pass rates?

3. Were there differences in pass rates between project and nonproject classes for students as a function of sex, ethnicity, or entering skill level?

4. Were there differences between project and nonproject classes in entering skill level composition, which might reflect a bias in student section selection associated with skill level?

5. How did students perform in subsequent mathematics classes (both project and nonproject) following enrollment in a project class?

While these questions certainly do not exhaust the potential information in the data collected, we believe they are the most important.

IMPLEMENTATION

During the period of the pilot program funded by NSF, data were collected for courses offered during four consecutive regular semesters: Fall 1988, Spring 1989, Fall 1989, and Spring 1990. Projects were used in a total of seven different courses: Math 142 (introduction to differential and integral calculus for biological and social sciences),

Math 242 (continuation of 142, with additional integral calculus, partial derivatives and differential equations), Math 191 (differential and integral calculus), Math 192 (antiderivatives and applications of integral calculus), Math 193 (an accelerated version of 191 and 192), Math 291 (infinite series, partial derivatives and multiple integrals), and Math 392 (differential equations). The courses and numbers of sections offered each semester were as follows:

| | | Number of Sections | |
| | | Project | Nonproject |
Semester	Course	Project	Nonproject
Fall 1988	Math 142	2	2
	Math 191	4	5
	Math 193	1	0
Spring 1989	Math 191	2	6
	Math 192	4	3
	Math 291	2	3
	Math 242	1	0
Fall 1989	Math 191	4	5
	Math 192	3	3
	Math 291	3	3
Spring 1990	Math 191	1	7
	Math 192	4	3
	Math 242	1	0
	Math 291	3	3
	Math 392	2	3

Average section size was 40 students, except for the single section courses (Math 193 and Math 242), which averaged less than 10 students per section. The sequence of Math 191, 192, and 291 is considered mainstream calculus designed primarily to satisfy the needs of students entering engineering, mathematics, and the physical sciences. We have focused on these courses for comparisons involving grade distributions, because they were offered in multiple sections over multiple semesters.

EVALUATION INSTRUMENTS AND PROTOCOL

Several forms of evaluation were used during the pilot study. The project investigators met weekly to discuss project concerns, much of which consisted of informal, on-going evaluation of the conduct of the program. These sessions frequently addressed details of implementation and planning for future improvements. In some cases, anecdotal evaluation was supplemented with data collected for specific purposes. For example, during the first two semesters of the program, detailed records were kept of lab usage by students. Analyses of these data allowed adaptation of the lab schedule to fit the observed patterns of student usage, resulting in some alteration to the scheduling of projects.

One major source of formal evaluation was the computerized record of students' course grades, previous mathematics enrollment history, and sex and ethnicity identifications, obtained from the university registrar. The Calculus Readiness Test (CR1) of the Mathematical Association of America (1979) was administered to students in first-semester calculus and used as a measure of entering mathematics skills. For subsequent courses, the grade earned in the prerequisite course was considered a measure of entering skill level. Efforts were made to obtain records of scores on common final exams in project and nonproject sections, but complete data were obtained during only one semester.

A second major source of formal evaluation was student self-reporting via questionnaires developed during the project. One questionnaire (see the Appendix at the end of this chapter) was given to students in classes near the end of each semester. This questionnaire included items concerning attitudes, interests, and study habits in mathematics, reasons for section selection, and, for those students enrolled in project sections, attitudes and activities associated with project work. Based on an analysis of responses from the first semester, the content of this questionnaire was changed and expanded for use in subsequent semesters; the summaries presented here do not include information from the first semester's questionnaires. Another questionnaire, administered at the departmental office to students withdrawing from courses, asked students to provide reasons for their withdrawal. Unfortunately, a large proportion of students who received W's did not initiate withdrawal, but were dropped by their deans on the recommendation of course instructors through reports of nonattendance. Thus, the withdrawal questionnaires collected at the departmental office generally represented only a small proportion of the total withdrawals. However, this did provide us with a sample of students' reasons for quitting the class when they did so voluntarily.

On the technical side, data from registrar records and CR1 scores were transferred electronically to a computer account maintained for the purposes of the project. These data were sorted and merged, and used as master archival data files for analyses. Each file included a header containing documentation of file contents and other relevant information, such as instructor names for each section, and designation of sections where projects were used. The basic information contained in each record included: social security number, last name, course, section number, grade in course, CR1 score (for first-semester students), sex, ethnicity, and major. Additional information collected by instructors, such as final exam scores, was also added to the archival files, when available. The statistical package SAS (SAS Institute, Inc., Version 5) was used to generate frequency summaries and statistical tests.

Grade distributions were calculated for two composite grade levels (Pass = A, B, and C, Fail = D, F, and W). These grade distributions were calculated for all students without subsetting, and for subsets of students identified by sex, ethnicity, and entering skill level. In the following descriptions, chi-square tests of independence were used in comparing pass rates in the various groupings of student grades, using a probability level of 0.05 for acceptance of differences as significant. Some descriptive comparisons are also noted for specific patterns of differences without reference to their inferential significance.

Data from class questionnaires were also coded and entered into computer files. Coding consisted of classifying responses into categories for each item by recording the category or the responses. Information for withdrawal questionnaires was summarized for review by the project investigators, but not computerized for analyses.

EVALUATION BASED ON GRADE DISTRIBUTIONS

As our pilot study progressed, we began to be overwhelmed by the complexity of patterns observed in our initial analyses of grade distributions, and overall generalizations appeared impossible. One pervasive pattern that emerged was a great deal of variation between semesters and among sections within semesters for every category of response in every course. At the end of Fall '89, investigators were elated at the high success levels of project students in Math 192 and Math 291, especially the indications of increased pass rates for females and minorities. This ended rapidly when the results for Spring '90 appeared, with a reversal in the overall pattern in Math 192, decreased pass rates for females and minorities in Math 291, and overall lower success for project students in Math 392. In addition to the factors previously cited as compromising project-nonproject comparisons, the progression of students through the curriculum, with their switching between project and nonproject sections from semester to semester, added another level of complexity in assessing outcomes in the more advanced courses. For example, when comparing the skill level compositions of Math 291 sections, should students who received a C in a project section of Math 192 be considered at the same skill level as students who received a C in a nonproject section of Math 192?

Given all of these sources of confounding, we have chosen to take a very conservative approach to interpreting our results. Based upon our comparisons with grade distributions of previous semesters, we cannot document that the initiation of the project option in our departmental curriculum resulted in any significant long-term change in overall pass rates for the courses involved (Table 1). In addition, all of the sectional pass rates observed during the course of the study were within the range of sectional pass rates from previous semesters.

Pass rates in project sections were definitely higher than in nonproject sections for some courses during some semesters, but Math 291 was the only course for which

Course	Pass Rate in Previous Sems.	Pass Rate in Current Study	Chi-square
Math 191	46% (1165)	46% (1263)	ns
Math 192	53% (657)	52% (733)	ns
Math 291	58% (577)	55% (597)	ns

TABLE 1
Comparison of overall pass rates in three courses before and during the study. Numbers in parentheses are sample sizes.

Course	Pass Rate in Project Sections	Pass Rate in Nonproject Sections	Chi-square
Math 191	46% (418)	46% (845)	ns
Math 192	50% (437)	53% (296)	ns
Math 291	62% (258)	51% (339)	$7.03, p < 0.01$

TABLE 2
Comparisons of pass rates between project and nonproject sections in three courses. Numbers in parentheses are sample sizes.

the pattern was consistent enough across all semesters to yield an overall significant difference (Table 2).

It is notable that for Math 192, pass rates differed significantly between project and nonproject sections for each semester, but project section pass rates were higher than nonproject sections for two semesters and lower in the third semester. Analyses of grade distributions using six grade levels provided support for one previously suspected pattern; there was a tendency for more failing students in project sections to drop the course (receiving W's) rather than staying in the class and receiving F's. There was little support for a suspected pattern of reduction in C's awarded in project classes (because "marginal" students tend to be pushed toward B's or D's), as percentages of C's awarded were similar across all courses, although percentages of B's and D's were generally slightly higher in project classes. Since our design for implementation of the project option is typical of what most institutions would find amenable, we are reassured by the overall results with respect to pass rates; we would be somewhat uncomfortable with results that showed extreme variations from the long term norm, because of the lack of rigorous experimental design that could provide us with identification of the exact mechanisms involved.

Attempts to analyze the success of subgroups of particular interest (females, minorities, low skill levels) were frustrated by sample sizes much smaller than the matching subgroups. Among the "majority" subgroups, the only significant differences were higher pass rates in project sections for males and nonminorities in Math 291, and for medium and high skill level students in Math 192. However, there was a general pattern of similar or higher overall pass rates among females and minorities in project sections as compared to nonproject sections. This suggests that there is some real potential that use of projects may be more effective than traditional methodology with these subgroups, assuming the implementation as used in this program.

We found no evidence of any significant overall differences between project and nonproject classes in the representation of entering skill levels. Although there were small differences among levels within each course during some semesters, these did not provide any clear pattern related to pass rates. For example, in Math 191 in Fall 1988, project sections had a higher proportion of low skill level students and a lower pass rate than nonproject sections. In contrast, in Math 191 in Fall 1989, project sections again had a higher proportion of low skill level students, but had a higher pass rate than nonproject sections. This agrees with the absence of a consistent pattern across all courses in the effect of projects on pass rates as a function of entering skill level.

Tracking student performance in subsequent mathematics courses as a function of previous project or nonproject enrollment yielded few consistent results. We did find, however, that for all cases except Math 191 retakes, students from projects classes who subsequently enrolled in project classes exhibited higher pass rates than students moving into project classes from previous nonproject enrollment. In two cases (Math 192 subsequent to Math 191, and Math 192 retakes), students who had projects classes had higher subsequent pass rates in both project and nonproject sections.

EVALUATION BASED ON STUDENT QUESTIONNAIRES

Questionnaires were administered during class sessions near the end of the semester. Student responses were anonymous. The timing of the surveys excluded students who withdrew from the classes and those who stopped attending but did not withdraw. Therefore, the samples represent primarily students who completed the course with an A, B, C, or D. However, the timing for the surveys was considered necessary, because classes typically had completed only a small part of the project experience by the last day to withdraw from the class. The following descriptions summarize survey results related directly to project mechanics (items #5–11, Appendix) and include responses from students in Math 191 (4 semesters), 192 (3 semesters), 291 (3 semesters) and 392

(1 semester). A more detailed report of complete survey results appears in Marsha Conley, et al. [3]

We found that many patterns of student response differed among the four courses, particularly in comparing Math 191 and Math 392. This may be partially a result of the small sample size (and single semester) basis of the data for Math 392. Of equal importance, student composition in each course also represents the result of progressive selective forces at work through the curriculum, both within mathematics and in other disciplines, with Math 191 and Math 392 at opposite ends of the continuum. Knowledge of the projects option had relatively minor influence on section selections in Math 191, 192 and 291, but a much greater role in section selection in Math 392. Since data from Math 392 were from the fourth semester of the program, a large proportion of students in this course had at least been enrolled in the sequence long enough to know about the option, and many had been enrolled in previous project classes. Overall, among students who had a definite preference, at least half expressed a positive response to the project experience, as demonstrated by indicating a willingness to enroll in future project classes. Most of those indicating a preference for future nonproject classes actually presented positive comments about projects, but objected to perceived greater time requirements of doing projects.

The survey was helpful in providing information about the actual mechanics of project work by students. Interactions with the instructor and the teaching assistants staffing the lab were of positive value to students, indicating that these should be considered essential elements in a successful projects program implementation. The content of negative responses to the TA lab indicated that students had not received adequate preparation in what to expect as assistance in the lab, and were therefore frustrated by the "minimalist" guidance by lab staff. It was notable that students were overall reticent and unsuccessful in using textual references for aid in project work. Typically, those who noted efforts at using the text or other references were not able to identify relevant information, although all projects were founded in extensions of basic course content, and many outside references were available in the library that would provide some useful information. It appears that students are not adept in identifying and using mathematical reference materials, which is not surprising, since this is not traditionally a standard part of mathematics courses. If this is to be a serious feature of a discovery experience for students, more specific planning is necessary to facilitate student efforts in using reference materials.

It was particularly instructive that students most frequently found organizational aspects (i.e., "getting started" and allocating time) to be the most difficult part of the project work, rather than the mathematics. Beyond the natural human propensity for procrastination, there appears to be a need to address this feature, perhaps by providing more structure to project assignments, such as intermediate progress deadlines.

---

[3] "Student Perceptions of Projects in Learning Calculus," to appear in *Journal of Mathematics Education in Science and Technology*.

Unfortunately, there is always the danger of introducing structure that detracts from the independent nature of the experience. Student enjoyment of the problem solving process was more common among more advanced students, while students in lower courses tended to focus on the satisfaction of project completion ("getting finished"). A decreasing frequency of negative comments in higher courses also supported the general scenario that higher acceptance of the project approach was exhibited by more advanced students.

The role of group learning versus individual learning as revealed in this survey was of interest in understanding the dynamics of the program, and is also relevant to general theories about learning style in mathematics. Although overall preferences for group work versus individual work were approximately equal, the trend across the course sequence suggested increasing facility and satisfaction with group work among more advanced students, as well as increasing flexibility for working in either setting. It is possible that this characteristic of learning style is a major feature of the academic success that allows students to reach advanced courses. Students' negative comments about group work most frequently referred to difficulty in scheduling group work and dissatisfaction with other group members' contributions. Difficulties with scheduling group work is an aspect of a project that would certainly vary with campus setting and student composition, but it is a feature deserving attention in planning the program. Responses indicating difficulty with group dynamics suggest that providing students with some preparatory information about group work might benefit the implementation of group projects.

In students' responses concerning their favorite project, the most notable pattern was that a large majority of students indicated a preference for projects that featured a "story line," with comments about relevance, applications, or "real life," even when the project setting was pure fantasy. The story line aspect of projects appears to be an important positive feature, which may function in initial visualization of the problem setting, or may simply function in stimulating the curiosity so essential to successful discovery learning.

In general, we found the results of the student surveys to be extremely helpful in obtaining input from students about attitudes and learning styles relevant to the project approach. We would strongly recommend that any plan to implement this sort of curriculum revision include provisions for acquisition, analysis, and dissemination of this kind of information.

EVALUATION UPDATE, SUMMER 1991
(Prepared by Gabriel Lampert)

University mathematics departments depend to some extent on the good will and co-operation of other university faculty and staff. By May of 1991, when I began work on evaluation of the project program, faculty members of several engineering departments had begun to question the program. Their claim was that the projects themselves took so much time that students (1) found it difficult to learn parts of the mathematics curriculum that were not immediately related to the projects, and (2) were therefore less prepared for the engineering courses they would later encounter.

The faculty members supported their claim with questionnaire surveys that had been given to engineering students. The questionnaires were not uniform in design, were sometimes unclear, and sometimes biased ("Do math classes give (a) too much (b) too little homework—choose one"). Despite these disqualifications, we could not ignore the fact that many of the students had listed (1) and (2) above in the "any-other-comments" section on the questionnaire.

We therefore focused this evaluation on the following questions:

1. **What effect, if any, does projects-class participation have on final exam score?** Math 191 and Math 192 have 'common' final exams: All sections of a course take the same exam. How well do projects sections fare in comparison with nonprojects sections?

2. **What effect, if any, does projects-class participation have on a student's grade in a concurrent engineering course?**

3. **What effect, if any, does projects-class participation have on the likelihood of student withdrawal from mathematics courses, when the student is concurrently enrolled in an engineering course?**

4. **What effect, if any, does projects-class participation have on a student's ultimate success in advanced mathematics or engineering courses?**

1. Final exam score. I hoped to obtain final exam scores for all Math 191 and 192 sections from Fall 1988 through Spring 1991. But faculty members keep their records each in his or her own way, so that some scores represent raw data and others follow a 'curve.' I tried to pick up information after the fact (instructors were not notified beforehand that their records would be needed in this way). The result is that many records were lost or not decipherable. However, there were data for the first two semesters, most of those scores having been recorded in 1989 by Marsha Conley.

For Fall 1988 (Math 191 only) and Spring 1989 (Math 191 and 192) sections, the mean final exam score for projects classes is higher in all three cases than the mean for nonprojects classes. For all three, the maximum possible score on the exam was 200:

		mean among all projects sections	mean among all nonprojects sections
Fall 1988	Math 191	134.58	120.26
Spring 1989	Math 191	134.95	120.98
Spring 1989	Math 192	125.02	116.24

It appears most unlikely that this difference is merely the result of chance selection. Instead of submitting these means to a t-test, I used approximate randomization,** which gave us these results:

		approximate p, using 10,000 iterations
Fall 1988	Math 191	.0168
Spring 1989	Math 191	.0243
Spring 1989	Math 192	.1356

2. Grades in concurrent engineering courses.   The single most consistent fact that emerged from this analysis was: The higher the level of the concurrent mathematics class, the more positive the effect of mathematics projects on the engineering grade. For most of the individual engineering classes, our sample sizes were too small to produce definitive results. The basic courses required by all engineering majors were the only ones for which our samples were large enough to put some confidence in our results. I also uncovered a startling fact: sometimes the concurrent mathematics class was a prerequisite for the engineering class. In this case the projects-sections students did more poorly, although almost all such students did badly in their engineering course. This is clearly not the fault of the projects program; it is an indication to those who would replicate the projects program at other universities that they might easily overlook the true causes for noted effects. In point of fact, there were few other significant results of any kind to this question.

3. Mathematics withdrawal rates during concurrent engineering enrollment.   Withdrawal from an enrolled course is an action chosen by the student, so it is impossible to know whether a student's choice to withdraw from a course is motivated by reluctant urgency on the one extreme or laziness on the other. And there are other confounding factors we cannot control for, such as student peer pressure (to withdraw or stay) and the student's adviser's collaboration. In any event, of the 11 engineering/mathematics

---

** Noreen, E. W., *Computer-Intensive Methods for Testing Hypotheses,* J. Wiley and Sons, 1989.

course pairs whose sample sizes were large enough to use, 3 showed an effect strong enough to note; in these 3, the withdrawal rates were always higher for projects-section students.

4. Student success in advanced mathematics or engineering courses. This is by far the most difficult sort of analysis to perform. To begin with, there are always more variables involved in student success rates than any we can gather data on. In addition, those variables we do have access to may act or interact in unknown ways. For example, if a student is forced to re-enroll in a calculus class (having gotten a D, F, I, or W), then will the first enrollment (project or nonproject) have an effect on the ultimate goal?

I decided to use the chi-square independence test, using two-way frequency tables. The variable of interest was the final grade in an engineering course (or Math 391, a higher-level mathematics course). In order to keep cell levels respectable, I kept 3 levels of grades: (1) A and B, (2) C, and (3) anything else. The other variables were zero/one (projects or not) variables for each of the mathematics classes which are prerequisites for the course of interest, and another variable which was the sum of the zero/one variables. Note that the zero/one variable was 0 if the student had enrolled in a nonprojects calculus class *the last time he or she enrolled in the course,* and 1 if that registration was for a projects course. Of those tables that were usable, most showed no significant relationship between projects classes and engineering grades. However, those results that were significant were all positive: students in some of the core engineering courses fared better in them if they had attended projects sections of Math 191, Math 192, Math 291, or some combination of these. The example given is for Mechanical Engineering 234 (Dynamics).

Observed ⟨Expected⟩		ME 234 grade			
		D/F/I/W	C	A/B	TOTAL
Math 192 class	nonproject	23 ⟨15.1⟩	24 ⟨20.2⟩	14 ⟨25.7⟩	61
	project	15 ⟨22.9⟩	27 ⟨30.8⟩	51 ⟨39.3⟩	93
	TOTAL	38	51	65	154

In general, it appears that there is a positive relationship between success in an engineering course and previous participation in projects calculus courses; in a few (important) cases, the relationship appears to be a strong one. In no case was there a strong negative relationship between projects-class background and success in the target class.

APPENDIX

QUESTIONS COMPOSING STUDENT SURVEY INSTRUMENT

1. What is the average number of hours per week that you spend outside of class on this course?

2. What has been the most valuable part of this course for you in learning calculus?

3. Are you more confident with your mathematical skills since you've taken this course?

4. Are you more interested in learning mathematics since taking this course?

5. How much time do you expect to spend working on a mathematics problem before you decide you can't solve it?

6. Of the many sections of this course offered this semester, what was your major consideration in choosing this section? Were there any other considerations?

7. Would you prefer a calculus course with projects again? Please explain your answer.

8a. What sources of help did you find most useful in doing the projects? Please explain.

8b. What sources did you find least useful? Please explain.

9. What was the most difficult part of doing your project?

10. What was the most enjoyable part of doing your project?

11. Which project type did you prefer—group or individual? Explain your answer.

12. Which project was your favorite? Identify by name. Give your reasons.

Chapter **3**

LOGISTICS: IDEAS FOR USING PROJECTS SUCCESSFULLY

Here we present specific information about using projects. We give guidance, and some unabashed advice, drawing on our own experiences and those of our collaborators over several years. Our intent is to tell you what has worked and what has not, so that using projects will be successful and rewarding for you and your students. We also want to help you avoid a few pitfalls along the way.

THE PROJECT EXPERIENCE FOR STUDENTS

What should the project experience be for a student? At the beginning of the term, students should learn about the role projects will play in the course. They should learn why the instructor believes projects are beneficial, including an indication of how they will help prepare students for the challenges encountered in future classes, the working world, and life generally. Projects should be portrayed as an exciting, natural part of learning calculus, not something strange or experimental. One student commented, "Projects give the student a more active role." In fact, students should be sufficiently encouraged by your description to feel that, while projects will be challenging, they will succeed at them, have a good chance of earning a good grade, and will feel a sense of accomplishment and self-confidence from the result, saying, "There is no better feeling in the whole world than finishing a lengthy project, turning it in, and knowing that you knew what was going on." In short, build your students' confidence that the experience will be worthwhile.

At the outset students should also learn a number of other things that set the stage for success. Particularly important are that the projects are mandatory, each worth a very significant part of the grade (typically 15–25% per project), and the assignments will be of limited duration. Students generally respond well to this: "It's great to have a problem to solve with a deadline—a real world experience—handling time." These three features appear to be crucial: one colleague at another university attempted to use some of our projects, but offered them as an optional alternative to the final exam, without limited time and firm deadlines. He felt his results were a disaster. While our students were successfully completing each project in the typical two-week period, his were not. Students were coming in for weekly appointments showing little or no progress, and he felt that he was doing the project for the students rather than the other way around. Since the only penalty for not completing the projects was taking the final exam, his students had insufficient motivation to devote the necessary time and intellectual discipline to the project. Just as with mathematical research, there must be a strong drive to go through the requisite agony for creating worthwhile mathematics. Two as yet untried possibilities for optional projects would be as an irrevocable option, e.g., the student chooses either projects or final exam early in the semester, with no possibility of switching, or as an alternative for poor test takers.

What else should students learn at the beginning of the course? They should learn how many projects there will be, which ones will be individual projects and which will be group projects, and that the projects are in effect replacing some in-class exams. If you choose to use group projects, students should be strongly encouraged to get to know others in the class to facilitate choosing partners for the group projects (this might be assisted by encouraging group homework papers). Our most common setup for a semester has consisted of one two-week individual project followed later by one or two two-week group projects, and we find we are moving more and more towards group rather than individual projects. It is important not to schedule an exam shortly after students complete a project, as studying for the exam will compete directly with the project, and one or both will lose. Assigning a project directly after an exam is fine. Typically the projects constitute 30–50% of the grade, with the balance from things like quizzes, homework, final exam, and in-class exams. Each project generally replaces what traditionally would have been an exam. Other possibilities we have tried include starting students in the first semester of calculus out with a one-week project or, at the other extreme, an all-projects course in which five two-week projects constitute almost the entire grade. In general we find that one-week projects are often rushed, and we have settled on two weeks as an optimal project duration. Almost all the projects in this book are two-week projects, with exceptions noted.

Now let us suppose you have reached the day when your first project will be handed to the students. In addition to giving them the project, impress upon them the time deadline for its completion. Convince them that if they wait until the second week to begin, disaster will ensue. Steady, diligent work is what leads to success. Give them a handout such as "How to work on your project; what is expected," or "How to work on your group project; what is expected," which are appended to this chapter. Perhaps give them a sample project, with corrected student mistakes in the 'solution' illustrating your expectations, the fact that the project is a writing assignment, and explaining what you expect if a project contains the words prove, show, or justify.

Let your students know they are not being thrown in the ocean to sink or swim, but rather are encouraged, and even expected, to get help from you and your project assistance laboratory (see below), if you have one. Explain that you want to help them avoid pitfalls and blind alleys. While students are working on their projects, you should be giving your class constant feedback on how they should be progressing towards completion of their project.

Tell them the rules about where they can seek help. Originally our students were allowed to talk only to their own instructor, but we have come to feel that this is both difficult to enforce and prevents constructive collegial interactions. Our rule now tends to be, "The paper you hand in must be your own work." Clearly some students collaborate or seek outside help, but we find they almost always learn more as a result. One thing that ensures that the work is the student's own is the advance expectation of a post-project interview (see below) at which the student will be asked to explain his or her solution. Originally we slaved to create and hand out multiple versions of each project to prevent collaboration, but we have found that this is not always effective or necessary. Several of us now feel that, especially with the role played by interviews, multiple versions are not worth the additional work.

BEHIND THE SCENES

There are several things that affect instructors more directly than they affect students: preparation and creation of projects, camaraderie, satisfaction and rewards for instructors, graduate assistant and undergraduate assistant apprenticeships, and benefits from the economy of scale.

Creating a project for use in the classroom is more involved than writing an examination. However, if you are using a project directly from this book, you may simply match a project with the timing of topics in your course. If you feel inspired to begin creating your own projects from scratch (as we hope you will), then you should allow two to three weeks for a project idea to gestate and mature into polished form. See our chapter "Creating Projects" for guidance. You will be amazed at how much fun creating projects can become; after you have done it a few times, ideas will come to mind everywhere and you will be surprised at how stimulating the mathematics can become. It is also useful and rewarding to create projects jointly with another instructor. Preparing projects is one place where instructors working as a team can benefit tremendously from economy of scale, and make the use of projects not a burden on their time. One thing cannot be emphasized too strongly about preparing a project for a class: *solve it completely before assigning it!* A typical project involves more quirks than instructors are likely to see at first glance, and you should avoid at all cost the disaster of running into serious problems with a project assignment after students are working on it.

Camaraderie and satisfaction will emerge not only from working together preparing projects, but from enriching interactions with your colleagues related to teaching.

You will also find your interactions with students more enjoyable. Your opinion of your students will rise as you talk with them about things they have studied in some depth. You will see their best work, unlike their in-class exams, which are often full of errors driven by anxiety about time.

ASSISTING STUDENTS

Assisting students with their projects requires extra time. It is the nature of projects that they are more intellectually demanding than the usual formula drill, and students need and deserve substantial guidance. But the benefit, that students learn how to think mathematically and *do* mathematics, makes it all worthwhile. Questions we have explored are how to make the time manageable by establishing a project assistance laboratory, how to benefit from economy of scale, and how to use different sources of assistance in order to take the primary burden off instructors. While most of our experience has been with using graduate assistants to staff the lab, we have also experimented with advanced undergraduates, and concluded that with very careful training they too should be able to staff the lab. The possibility of using advanced undergraduates to provide lab help is particularly important for those institutions that either have no graduate assistants available, or have foreign graduate assistants with inadequate English communication skills, especially since the interpretation of English is more important for projects classes than in typical mathematics courses. The lab also provides an unusual opportunity for a true teaching apprenticeship in association with experienced teachers, since close interaction about mathematics and pedagogy occurs naturally among those involved.

It is crucial to realize that the purpose of the laboratory is not to solve the project for the student. On the contrary, it is to support the student in doing the project himself or herself by giving encouragement and providing just enough help to get over hurdles to enable further independent work to occur. This idea is perhaps a bit alien to us, since when a student comes to our office for help, we often say, "OK, here's how you do it", and send the student on his or her way again. In the lab a question is just as likely to be answered by another carefully chosen question, or the assignment of a supplementary exercise, the aim being to help the student ask the questions that will lead to self-discovery. This is one of the reasons why training, supervision, and consistency are extremely important for those who will staff the lab. We have developed a set of "Guidelines for People Working in the Projects Lab" and "Instructions to Teaching Assistants" (appended to this chapter) giving general expectations and guidance. Before each project assignment begins, all those involved must meet to go through the project in detail, discussing possible pitfalls and agreeing on the appropriate level of hints. It is important that those staffing the lab provide guidance rather than do the work for the students. Instructors and lab staffers should also discuss and refine their approach while the lab is underway, based on laboratory experiences.

In the next few paragraphs, we will explain the scheduling and staffing logistics for the lab as it works for us, and you can adjust for your own situation. Suppose we are running a lab for four 40-student 3-credit lecture classes of the same course. These classes have no separate problem and recitation hour, and simply meet 150 minutes per week. The classes have the same project assignment dates, so that all 160 students coming to the lab are working on the same project with the same due date. All four classes will use the same lab. We have arranged for the instructors of the four classes to consist of two faculty and two graduate assistants. Traditionally each graduate assistant would teach two classes. However, those teaching in our projects program will teach only one class and staff the projects lab for the 160 students in these four classes. Well trained advanced undergraduates for lab staffing could partially substitute for staffing by graduate assistants.

During a two-week project assignment, we expect each graduate assistant to staff the lab for 16 hours. This occurs two or three times during the semester. Since the lab is of a drop-in nature, its staffing is tailored to student demand, which will increase steadily during the assignment period. The total number of hours of actual lab staffing that we expect per graduate assistant is not exorbitant, but it is heavily concentrated into just a few weeks when projects are assigned.

As an example, suppose a two-week project is assigned on a Friday in a second semester course. Since most students have classes and labs in the morning and early afternoon, the best times to schedule the lab are late afternoon and early evening. Our lab will be staffed by one graduate assistant for two hours daily on the first Monday and Tuesday afternoon, and four hours daily on Wednesday and Thursday in the afternoon and early evening. On Friday we double staffing levels to have both graduate assistants on duty simultaneously. With more than one lab staffer on duty, separate rooms help ease congestion. However, the Friday lab runs only during the two afternoon hours, since we have found that on Friday evening the calculus lab is not always an entertaining nightspot. We have sometimes provided weekend lab times. The lab continues doubly staffed for four hours daily on Monday and Tuesday, and possibly Wednesday.

Even though these projects were not due until Friday, the last lab day was Tuesday or Wednesday. Closing the lab early encourages students to complete most of their work in a timely fashion. If a student has made poor progress with the project up until a day or two before it is due, and is somewhat desperate, the situation should be dealt with delicately by the student's own instructor. This avoids unpleasant and unreasonable demands on lab staffers, and allows the instructor to take into account his or her fuller knowledge of the student's background in dealing with problems, and to make special arrangements if appropriate. During the final day or two, instructors should have extended office hours, while emphasizing in advance to the class that not all students can receive special help in such a brief period. The interactions during this last day or two are often very rewarding, as many excited students come in with almost completed projects, for which they either want to clear up a detail, or to have you look over a draft.

At the lab itself, students simply come forward individually for help from the staff. If there is a wait, a sign-up list on the board obviates forming a line. Rules limiting individual help to 5 minutes at a time can be helpful in periods of high demand. You may wish to monitor use levels to adjust your staffing schedule, perhaps by adding supplementary undergraduate staffing. It is important to note that we have also found that students in first semester calculus seem to need more help than this schedule provides, while in later semesters they can do with less assistance. We tried an optional lab by appointment in response to complaints about having to wait, but it met with little interest, with students preferring the drop-in method. Finally, note that if some of your classes meet only twice a week on Tuesday and Thursday, the same lab schedule can be used to serve both types of classes by simply assigning projects on Thursday and Friday.

GRADING AND INTERVIEWS

Private 5 to 15 minute interviews with selected students after projects are completed can play several important roles, the first two of which involve grading. The interview helps refine a tentative grade obtained by reading the project, and sometimes influences that grade substantially. Clearly the interview (or simply the possibility that one may occur) also works to ensure that each student's work is his or her own. Generally we have interviewed every student after the first project of the semester, but only selectively thereafter. Each student nevertheless expects the possibility of an interview on any project. One way to reinforce this is as follows: hand back graded projects, telling the class that you have written a note on the projects of those students who are to see you privately for interviews. The class has no way of knowing how many, if any, students have received such instructions. You can adjust your use of interviews to your own needs.

The post-project interview also has great pedagogical value. Students gain a lot from discussing and explaining their project, and obtaining the instructor's perspective in person. One student said, "Students had the opportunity to explain verbally what they could not explain in written form." They can also be advised and counseled about improving their work. Most of those who have done poorly recognize the reasons why (commonly procrastination), and correct these things for the next assignment. Finally, the interview presents a sterling opportunity for all of us to encourage good students to pursue and perhaps major in mathematics.

Some of us return the marked project before the interview, expecting the student to address marked issues at the interview. Others of us do not, preferring to address issues first during the interview. Avoid doing most of the talking, and expect students to explain things instead. The final grade is only assigned after the interview. We experience very few student objections to their grades.

Generally the project grade is a subjective letter grade, just like an English paper, although a few of us still use a points system. Most of us find that, with some practice, it takes less time to read and grade a project than an exam. It is a genuine pleasure to read and grade students' projects, as opposed to exams. We are seeing their best work, and not their worst, and are able to see the fruits of their labors. Moreover, one can often see how they have mastered ideas they were only recently struggling with, corrected serious misconceptions (exams don't usually have this effect), and gained pride and satisfaction in their accomplishments. Students often say things like, "Most enjoyable. . ... Actually seeing yourself accomplish something that appeared impossible at the beginning." One feels they have earned the good grades they usually receive. It is a delight to see projects embellished with story lines and artwork.

Projects also seem to force students to face up to reality in a way exams often do not. Students may find excuses for poor performance on exams, whereas failure to do well on a project generally prompts serious self-examination. Some of these students make dramatic turnarounds to succeed on future projects and in the course, and even credit projects with changing their attitude about mathematics and themselves. The rest usually withdraw from the course after realizing they will not complete the projects, leaving very few failing students remaining in the course. We generally feel that those who withdraw are amongst the students who would have failed the course with or without projects. We always arrange to have a project graded and returned before the last day to withdraw (midsemester at our university).

GROUP PROJECTS

We discovered that having students work on their projects in groups of two or three can have tremendous advantages for both students and instructors. Consequently group projects have become a standard part of our program.

For students there are two advantages to group projects. Working in groups is probably a more realistic approximation to the environment they will enter after graduation, and thus group projects teach them useful skills. Students themselves believe this quite strongly. Many students are also more highly motivated in the social context of group endeavor. One student commented, "I feel more confident taking a class where I don't feel like I'm alone."

The principal advantage for instructors is that students require considerably less assistance, since they are learning from each other, and each group turns in only one paper, resulting in less grading. The reduction in assistance needed by students is quite noticeable, and can result in lower lab staffing levels and fewer extended office hours by instructors. These savings help make the program practical for instructors. We strongly recommend that you incorporate some group work in your use of projects.

There are some disadvantages to group projects as well. First, while many students like group work, some don't, although the balance of opinion seems generally favorable. Second, students normally have little experience in collaborative work.

Training students to work productively in groups is important, although we have not yet concentrated enough on this. Third, a potential difficulty is conflicts between the schedules of the students in a group, preventing the group from meeting regularly. Students should be instructed to take their schedules into account when forming groups, so that daily meetings are possible.

Perhaps the greatest worry about using group projects is that it might make evaluation of individual students' work difficult, and that weak students will receive a free ride. While these are issues of concern, we find that the benefits far outweigh the disadvantages, and that there are relatively simple ways of dealing with these problems. At the beginning of the group project assignment, students are told that when the project is completed, each student will also give the instructor a private, signed personal evaluation of the role each individual in the group has played in the completion of the project. In its simplest form, this merely consists of a statement of the percentage of work each member of the group contributed. It could be expanded to a paragraph describing the group's dynamics. Our experience is that the group's members almost always agree in their percentages, even when they show that one student did less than others. In fact, if anything individuals tend to judge themselves more harshly than their partners do. Dilemmas over disparate estimates of work are rare, and we use interviews to clear up discrepancies when they arise. We do assign different grades on a project to different members of a group if it is clear that one student did substantially more or less than others. We should mention that the group projects we assign are often more challenging than individual projects.

When preparing your class for a group assignment, have them get to know one another early in the semester, possibly via group homework, and insist that they form groups and set up meeting times a week or two before the project is assigned, so that this does not cause trouble once projects are handed out. You might consider having them get started on the group project for 10 to 15 minutes in class on the first or second day. If individuals have strong reasons why group work is impractical, e.g., long distance commuters or serious schedule conflicts, then we allow them to work alone. Occasionally you may have to alter or dissolve groups if special problems arise. Finally, you may wish to set up special rules for groups regarding assistance. For example, in the lab we usually insist that only one group member come for assistance at a time, and report back to the group. This has the pedagogical purpose of forcing the student to explain what was learned in the lab to his or her colleagues, and can also relieve lab congestion. On the other hand, working with the whole group reveals its dynamics to the instructor, which is also useful and often enjoyable.

INSTITUTIONAL SUPPORT

Clearly for a projects program to be successful on a broad scale, institutional support and economy of scale will be important. Perhaps the most obvious way in which institutional support may be important is to provide the human resources for staffing a projects lab, using some combination of graduate assistants and advanced undergraduates. If these apprentices are to do justice to the intellectually demanding task of helping students with difficult projects, they must be carefully trained and supervised by an experienced faculty member. In our university, our department faculty, department head, and dean have seen that this use of resources is warranted by the benefits that derive for our calculus program.

In addition to the allocation of lab staffing resources, we have a faculty member released one-quarter time to serve as "Projects Program Coordinator," to handle the many aspects of running a successful program. This includes, for instance, recruiting and training new volunteers as instructors (teaching projects courses is strictly voluntary), training and supervising lab staffers, assisting with the creation and preparation of projects, scheduling individual and group projects, assignment dates and lab staffing, and serving as a liaison with other departments and student advisors, since we now offer a mix of both project and nonproject courses.

APPENDIX A

HOW TO WORK ON YOUR PROJECT: WHAT IS EXPECTED

Here are some words of advice and encouragement along with the rules of the game.

First of all, this is a major, lengthy assignment. To do well you should start immediately, and work on it every day. You will probably need all the days you have been given in order to complete your project by the day it is due.

1. Start today. Let your subconscious work for you. It can do amazing things. If you immerse yourself in the project, solutions will come to you at the strangest times.

2. Read the entire project to see what it's all about. Don't worry too much about details the first time through. Do this today.

3. Next, read the project very carefully and make a list of any unfamiliar words or concepts you encounter. If concepts occur that you're not sure about, you must understand those ideas before you can do the project. Even if you understand all the words and terms, don't assume that the project is easy. If you wait until the last few days or so to start, it is doubtful you will be able to finish on time.

4. You may need to do some outside reading. In addition to your textbook, there are lots of books in the library that contain information that might be helpful to you.

5. After you have worked a bit every day on the project, you will find certain parts easy and you will have completed those parts. You will have identified the hard parts and have begun to zero in on the obstacles. You are beginning to become the master of your project. You are so familiar with it that it is easy to sit and work on it if you have a spare minute or two. I recommend you work on the project some every day and keep a journal to record your progress.

6. While I expect you to work independently, I do not expect that you can work through the project without some assistance. I encourage you to come and talk to me about your project, and to get help in the special Projects Lab for your course. Even if you think things are going along smoothly, you should let me see what you are doing. This way I can head you off if you are going in a wrong direction.

7. There will be times when you need help and I'm not available. At those times, you should go to the Projects Lab for assistance. It is open several hours daily for most of the days you will be working on your project. You will receive a schedule showing the Lab hours.

8. Don't go to the Projects Lab expecting someone to work your project for you. The purpose of the lab is to give you guidance and let you know if you are on the right track. Come with specific questions, and be prepared to show clear written work that you have prepared in advance.

9. When you have done the work necessary to complete the project, you need to prepare it in written form. The paper you turn in should have a mix of equations, formulas and prose to support your conclusions. Use complete sentences, good grammar and correct punctuation. Spelling is also important. The prose should be written in order to convey to the reader an explanation of what you have done. It should be written in such a way that it can be read and understood by anyone who knows the material in this course. You will be graded on your written presentation as well as the mathematical content.

APPENDIX B

HOW TO WORK ON YOUR GROUP PROJECT: WHAT IS EXPECTED

This project is due in two weeks. Remember, it is a major assignment, so you will want to do well. This is a group project and I offer some suggestions for group work.

1. You should plan your first group meeting as soon as possible, but before that first meeting, you should have read the project carefully and given it some thought. At your first meeting, you should plan a method of attack and you may wish to divide the labor among the group. Different members of the group may perform different tasks, but all members of the group are expected to understand all parts of the solution. Regular group meetings to discuss the progress of the solution are important.

2. You may seek assistance in the Projects lab or from me. The schedule for the lab is attached. You would be wise to consult with me or the person in the lab, even if you think things are going well. This will help you avoid dead-ends.

3. The paper you turn in should have a mix of equations, formulas, and prose to support your conclusions. The prose should be written in complete sentences that convey to the reader an explanation of what you have done. It is not necessary that projects be typed, but if handwritten, they should be neat and legible. Each group will turn in one paper.

4. You may not use the Projects Lab as a place for your group to come and work. If you have questions, you should send one representative to the lab for help. Then that person can report back to the group. I would appreciate the same courtesy for visits to my office.

5. At the time the projects are turned in, you will hand in a sheet giving your assessment of the relative contributions of yourself and the other members of your group. It is not automatic that all members of a group will get the same grade on the project.

APPENDIX C

GUIDELINES FOR PEOPLE WORKING IN THE PROJECTS LAB

1. Our goal is to challenge the student to think. Whenever possible, answer a question with a question, but in a way which gives the student a direction for concrete progress, rather than frustration. This challenges students to question their own reasoning and hopefully instills the confidence needed to solve the problem.

2. A student will depend upon you to provide help at every step, and our purpose is to gently wean them from this. One method of giving stingy hints is to write nothing down for the student, but rather keep the question and answer session verbal. Consider leaving your pencil at home. Even better, demand the student do the writing.

3. Encourage specific questions from the student. Above all, do not let the student do work at the desk where you are sitting. Send such a student away to think about what you have said. Set specific time limits for your encounters with students. No individual encounter should last over five to ten minutes.

4. A frequently asked question is "Where do I go from here?" or "What do I do next?" The best method of dealing with these questions is to know the "holistic" structure of the project. Oftentimes, students have no rhyme or reason for what they are doing, and you may have to motivate the whole project. Give the students reasons for what they are doing. The reasons can serve as guideposts for the student without giving away any answers, and usually quell their anxieties about what to do next.

APPENDIX D

INSTRUCTIONS TO TEACHING ASSISTANTS

The purpose of this document is to describe our philosophy about your role and obligations regarding calculus projects.

You have a dual role in this department. You are both a student and a teacher. It is important for all of us not to lose sight of this duality and to keep the two roles in perspective. It is also important for you to know that we as faculty recognize the importance of both roles. We do not want your teaching to demand so much of your time that your graduate studies suffer. On the other hand, you must be willing to devote the time necessary to meet your responsibilities to your students.

As a teaching assistant involved in using projects, you have duties that differ somewhat from those of other teaching assistants. Some of those are detailed below.

1. You must meet with your course coordinator before the semester begins. Details, such as the number of projects, dates for projects, number of examinations, homework, etc., must be discussed prior to the first class meeting. Your syllabus should be designed in consultation with the course coordinator and should include pertinent information relative to projects and examinations. You should know enough about the use of projects to field related questions on the first day of classes.

2. At least a week before projects are assigned, you will receive copies of the projects for you to solve. In order for you to be able to help students with the project, you must understand it yourself, and the only way to know the project is for you to work out a complete solution. You will then participate in a preparation meeting with all those involved before the projects are handed out.

3. During the period that the projects are out, classes should proceed in the normal manner. Homework should continue as usual. There may be occasions where it is appropriate to spend part of a lecture on a topic directly related to the project.

4. You will be assigned times to work in the lab for project students. This is an important task and you should plan to be prompt and prepared for the lab. You will be expected to keep a head count at the lab. Your course coordinator will have discussed with you the types of hints that you can give, since you will not actually solve the project for the students. An important part of your job in the lab will be to help students understand what is expected and keep them from heading down blind alleys. Do not agree to proofread a project for someone else's student. This can cause problems.

5. There will be a few days between the last day that the lab will be open and the day the project will be due. During those days, you are expected to have extended office hours for the purpose of helping the students in your section. Be careful that

students don't camp in your office and work on their projects. You might post a sheet and have students sign up for a time (perhaps at 10 minute intervals) to avoid a line in the hall.

6. After the projects are turned in, you should plan to read them and return them promptly. The next class period is ideal; don't let more than one class period pass before you return the papers.

7. Read the project carefully, mark errors, note bad grammar and incorrect use of language and notation. If the errors in language and notation are excessive, make a note on the paper to that effect and don't try to mark all the errors. You may want to correct some of the notation and language errors, but don't correct the mathematics.

8. Discuss with your course coordinator which students you should interview, and ask these students to schedule interviews. Some of us return marked papers (without grades) to these students prior to the interview, while some of us retain the papers until the interview. All other projects should be returned with grades indicated. Grades for students to be interviewed will be assigned after the interview and graded papers should be returned to the student at the first class meeting after the interview if at all possible. Be prepared to discuss the grade with the student if requested. It is not uncommon for students to expect good grades on mediocre papers.

9. After each project cycle is completed, be prepared to give your course coordinator your impressions and comments.

Chapter **4**

CREATING PROJECTS

Here are some hints and suggestions for creating your own projects. There is advice on where to seek ideas, on how to turn these into full-fledged projects, on testing them to eliminate blunders, and on tying them into specific courses.

Why should anyone need to create their own projects? Why would anyone want to? Well there are only about one hundred in this book. After a few years these will be exhausted, unless one wants to repeat them with the risk of solutions resurfacing from fraternity files. Even though there are many excellent collections of mathematical problems, they are not in a form, generally speaking, that is immediately usable as projects in a calculus course. Thus they need to be adapted. Finally, one may want something centered around a particular topic not represented in our collection.

As to the second question, it is fun to think up problems. It's satisfying to take a primitive impression and expand and shape it into a full-blown project. The creative satisfaction of developing a project is somewhere between constructing the harder homework problems in a traditional course and doing light mathematical research. The effort to create a project from scratch, vet, and edit it into final form, as seen in this book, is significant but usually less than that of writing a short note for the *American Mathematical Monthly*.

SOURCES OF IDEAS

How do we get started? Where can one find inspiration for student research projects? Well, ideas for them are everywhere. You can get some out of books, but we tended to eschew amplifying standard problems. We preferred creating our own. But you first might want to consider extensions of problems in textbooks. Sometimes a student will ask a probing question in class and this will lead to a new notion. You can look in the mathematical literature under problem collections and in problem sections of journals. Consult people, books and journals in other disciplines. Journals meant for teachers in all disciplines are especially useful. Be alert to situations in everyday life. You can even have students create problems. In a moment we detail each of these possibilities for the reader eager to uncover rough nuggets that can be refined into shiny new projects.

When we first started compiling this collection, we thought that we would shortly run out of ideas. We mistakenly thought there was a very limited pool of concepts that could be turned into workable projects. How wrong we were. Our current view is that the supply is virtually inexhaustible. The limiting factor turns out to be the labor required to turn an initial conception into a fully written and completed enterprise, with the solution fully explored to make sure that it is within reach of the intended audience.

The obvious place to start looking is in one's textbook. There may be problems that can be enhanced to make them more difficult: higher dimensional analogs, more parameters, harder functions to work with. For example, in this book there are four projects on tangents: two in two dimensions, *Equal areas* (#23) and *Invariant areas* (#24), and two more in three dimensions, *Tetrahedra* (#78) and *Tangent planes to cones* (#81). One can expect students to extend the theory, create counterexamples, etc. While this approach sometimes yields good results, other approaches are often more fertile.

Projects may be used to enhance the learning of material that students typically find difficult. At our school, one such topic is vector algebra because of its placement in our calculus sequence. Poor student performance on this material prompted us to write challenging questions related to vectors, vector algebra, and the relation between vectors and calculus. One such project, *Escape from the Cleangons* (#74), involved the Starship Exitprize being held captive in an elliptical orbit by the Cleangons. The students enjoyed this and many entered into the spirit of it with their own story line. Some students even took the time to learn about the equations of ellipses in polar coordinates, which are not covered in class, and used them to solve the problem.

Projects may be used to insert important material that is not included in the course. There have been proposals in recent years to eliminate certain techniques of integration from the calculus course. Such topics can be made part of a project. Two of our projects have involved the use of hyperbolic functions as a technique of integration. Students were asked to read about these functions and then discover and prove the identities necessary to make hyperbolic substitutions useful.

There are many collections of problems in the literature, e.g., Albert A. Blank, *Problems in Calculus and Analysis*, Wiley, 1966; G. N. Berman, *A Collection of Problems on a Course of Mathematical Analysis*, MacMillan, 1965; Murray S. Klamkin, *International Mathematical Olympiads, 1978–1985, and Forty Supplementary Problems*, Mathematical Association of America, 1986; Richard C. Allen, Jr., and G. Milton Wing, *Problems for a Computer-oriented Course*, Prentice-Hall, 1973. Some may be too easy and others may appear too advanced, but with suitable reworking and perhaps hints added, all these books offer good ideas for projects. Many journals have problem sections, e.g., the *American Mathematical Monthly, College Mathematics Journal*, and *Physics Education*. The last is in a field other than mathematics and published in the United Kingdom, but nevertheless some of the conundrums posed are suitable for calculus courses.

This demonstrates the value of looking abroad and in educational journals in other fields, e.g., see in *Physics Education*, 'The great water-jet scandal,' **23** (1988), pp. 137–78, 190–91 and **24** (1989), pp. 7, 67–68, which is a delightful, witty sequence of notes debating the nonintuitive distances that water spouts out of a can from holes punched at different heights. Typically, problems found in journals will not be immediately usable. However with an appropriate creative flair, one can refashion them to one's taste. If you have access to a large library with older holdings, consult such stimulating series as *Mathematical Questions and Solutions* in *The Educational Times*, after the turn of the century, and *Encyklopädie der mathematischen Wissenschaften mit Einschluss ihrer Anwendungen*, around the turn of the century. This list is not meant to be exhaustive, but merely indicates how imaginative browsing in a good scholarly library will turn up all sorts of possibilities for worthwhile problems.

Do not overlook valuable resources: colleagues and journals in related disciplines such as physics, electrical engineering, computer science, and biology. Despite this exhortation, we did not seek out people in other departments as avidly as we are advising you to do. We did concoct some projects by working with colleagues in biology and computer science, and received quite a few clever ideas from persons in physics, chemistry, and electrical engineering. But the latter were never turned into full fledged projects for lack of time and energy. Despite the rewards of cross-disciplinary collaboration, both for faculty and students, it does take considerably more effort than working strictly within a mathematics department. Anyone contemplating this source of potential projects must be willing to learn to talk to people in disciplines outside mathematics.

Situations in everyday life can stimulate the mathematical mind and lead to delightful assignments that students love. You can get ideas while walking across campus, or at home, or even while dining out. Often there is an extraordinary change of emphasis from the original flash of inspiration in your brain to the fully illuminated manuscript you hand out to students, with twists and turns of imagination not easily anticipated. For example, if you go to a pizzeria and order a pizza, you typically see sliced olives on it. These are little disks with the center removed where the pit was. This image led to a very successful project, *Jacobi's Pizzeria* (#100). If you take a sphere and you have two blades that are always at a fixed distance and you put the sphere symmetrically through them and slice off a couple of caps, the caps will be circular. If you further remove a cylinder from your olive whose cross-section is exactly the size of that circular disk, then you'll end up with a pitted olive. What's amazing is that the volume left in your pitted olive is independent of the original size of the olive, assuming it's big enough so that there's something to slice off. We had students prove that and then we asked them: "Olives aren't spheres. They're more like spheroids. If you start looking at various shapes other than spheres, does this result still hold?" This in turn led to a study of the Jacobian in order to perform the integrations so that one could decide for which shapes of olives this would be true. One student actually reported a breakthrough after having dinner at a local pizzeria.

Barbara White, at the University of California at Berkeley, has suggested encouraging or even requiring students to think up advanced questions. This is not as crazy as it sounds. To develop a project often requires the author to pursue a number of different lines, involving actually solving more than the final product will require. We sometimes find ourselves working with more parameters than needed just to find out the best numbers for the final version. Thus any student seriously designing a project will get a good mathematical workout pursuing both fruitless and profitable lines of investigation.

VARIETY

The range and depth of projects that can be created from one initial notion is large, much more than one would expect. After one has a bright idea, one should next seriously entertain what length, style and pedagogical motives are desired. We list a few of the possibilities:

> minimalist versus road-map
> spatial versus strictly logical
> pure versus applied
> computer versus brain
> inductive versus deductive
> short versus long

We now describe in detail some of these different styles, citing advantages and disadvantages.

The first dichotomy concerns what is sometimes called road-map versus minimalist: detailed instructions versus a short succinct description and summary of what is to be found. Oftentimes the minimalist project involves neither numbers nor symbols. Needless to say, students initially are overwhelmed, then go on to sort out the challenges and crack the project with a good deal of satisfaction. On the other hand, some of our project statements tend more toward the "road-map" approach, in which a student's progress through the project is guided step-by-step. Similar observations may be made about posing broad questions versus listing detailed instructions.

A Greenhouse extension (#31) is a good example of one of our minimalist statements: only the data essential to setting up a well-posed problem are given, and those are given only in words without symbols or numbers, and with no clues as to how to turn that data into mathematical expressions. Instructors may expect to have to field many questions from students as they first get started with this type of project.

Students generally like assignments that they can visualize. Story lines exploiting geometry are quite popular. Hence several of us have gone out of our way to create imaginative characters and clever dialogue, or adapt those popular in television or cartoons.

Harder for most students and less enjoyable for some are abstract problems with lots of algebra, or worse yet logic. Balanced projects often have both inductive and deductive reasoning. Having both modes of reasoning in the same project forces students to confront their differences and compare their values.

While perusing the projects in the second half of this book, you will notice a number that might be perceived as applied. Do not be deceived, for one should imagine a continuum of problem types from pristine pure mathematics to dirty applied mathematics. For purposes of discussion, we recognize four benchmarks: pure, fake-applied, pseudo-applied and applied. Pure mathematics we all know and recognize. At the other end of the spectrum, an applied problem is the application of mathematics to a specific problem in science or engineering: typically these are too hard and involved, and too intimately connected with a specific branch of knowledge outside mathematics, to be suitable for our purposes. A pseudo-applied problem, on the other hand, is obtained from a truly applied problem by simplifying the assumptions, stating the necessary concepts required from science or engineering, and keeping all this to a minimum consistent with the spirit or essence of the original problem. With an understanding of both mathematics and the field of application, and a considerable effort, one can construct attractive cross-disciplinary projects in this way. Unfortunately, very few of ours fall into this class. Finally, by fake-applied we mean projects that appear to be pseudo-applied, especially to the student, but which turn out to be unrealistic when examined closely. An example is *Escape from the Cleangons* (#74), cited earlier; this is not really where a spaceship should fire its engines. Such projects delight students and they think they are doing real science and engineering. Is this good or bad?

Some of our projects require a pocket calculator, e.g., *The logistic equation* (#102), and a few need a computer to solve, e.g., *The logistic equation with delay* (#103). However, our goal was to develop a program that could be used to enhance the understanding of calculus without requiring large financial outlays for equipment and software, and future commitments to maintenance and eventual replacement. Many more projects could be created for calculators and computers.

As mentioned elsewhere our assignments may be one-week, two-week or even longer. The same problem can be refashioned to make it easier or harder. For example, measuring an irregular area comes in two versions, one for the whole body, *Finding your area* (#36), and one for just a part of it, *Finding the area of your hand* (#37). Difficult projects can be made easier with hints, more lab help, handouts or even some discussion in class.

Perusing the projects themselves, the astute reader will notice only one picture. The primary reason is that we believe students should learn to draw diagrams from verbal descriptions.

STEPS IN CREATION

Once we have the kernel of a project and have decided on thc basic style and form of the project to be developed, we next decide what kind of mathematics we want the students to do. What objectives do we place before the students and what tools are they expected to use? With these points in mind we go ahead and write a first draft.

Now is the time to get a second opinion and keep getting second opinions until the project is finished. In fact, in our group we would always designate someone as a secondary editor to read over each new version and comment on it. This person's duties included carefully working the problem through from the standpoint of the student to make sure there are no howlers. The primary editor was usually the person who thought up the idea first. Incidentally, in preparation for publication, all five of us read over each project at least once, although not all of us worked through the details of every project.

The primary and secondary editors now rework and revise in the light of back-and-forth discussion. Often the viewpoint changes considerably, as the two editors share concerns and possible extensions. As a case in point, one project started out life by asking students to find the average distance from the center to any point within a hexagon. Then one of us pointed out that there was nothing special about a hexagon, it could just as well be done for any regular polygon. Finally, a story line was added to give us the final version appearing in the collection, *Planning a community* (#92).

Next the project is carefully written and edited. Care must be taken to make sure that the statement of the problem is unambiguous, that it can be solved with the tools the student has, and that the effort required is reasonable. Although errors can be corrected after the students have a project, that can be exceedingly awkward, especially in multisection courses. So complete vetting is strongly advised. And, of course, since you will be grading your students in part on how well they write and explain their solution, be sure your statement of the problem is in correct English.

Allow enough time to go through all these steps so that copies can be run off in time to be handed out on schedule. Even with experience, writing a good project takes more time and preparation than an exam. Without prior experience you should probably allow four weeks from the time you first conceive the kernel to when the polished product is handed out. Not that you will be spending all your time on writing and editing during these four weeks, but that the primary and secondary editors need leisure time for the concept to gestate. Also, if you are using graduate assistants or undergraduates to give help to students, then these people will have to be thoroughly introduced to the enterprise ahead of time and made aware of any special instructions.

You may wish to add some boiler plate text outlining what the students may expect in the way of help and a laboratory, if there is one, as well as what you expect in the way of a write-up from them. And now, as you are about to hand out the projects to the students, you should turn to the chapter "Logistics: Ideas for Using Projects Successfully."

Finally, after you have listened to questions and the students have been helped, they turn in the assignment. Now is the time to reflect on unforeseen difficulties and hang-ups, or even where things might have been too easy. In light of your experiences with students, edit the project one last time, so that it will be ready for use again in the distant future.

Despite a team's most conscientious effort at creating a viable problem, sometimes it will be too hard and flop. For example, *Measuring the USO* (#75) did not work out when put before students. They found it too difficult and incomprehensible. We thought it would be a good project, but it just didn't pan out.

OTHER POINTS

How closely should a project be tied to the material being covered in the course at the time the project is assigned? There is a wide range of opinions on this. One extreme is not to try at all, the attitude being that one is teaching mathematics and how to do it, and therefore what the topics are is immaterial, so long as students are prepared to tackle the project. In other words, according to this view projects need not be timely to be effective. For example, at the beginning of the first semester of calculus, we have assignments only requiring high school mathematics with no calculus; these are valuable for uncovering and repairing defective knowledge. The other extreme is to have students do a project at exactly the same time that material essential to its solution is being covered in class. This sure perks up the students, making them very attentive and serious in class. Unfortunately, it is hard to pull this off well, especially at the same time in multisection courses.

Likewise, one member of our group is experimenting with writing collateral material that dovetails with a current project. As a result, students are picking up new concepts by discovering for themselves their utility for the project at hand. This requires the posing of leading questions and examples, both in the notes and during class.

There is another philosophy of adding parts for extra credit beyond what is expected of all students. This has evolved since we began writing projects. Some of us started posing parts for extra credit. This was picked up by the rest of us with the result that now a fair number of projects have extra credit questions. As remarked elsewhere in this book, these extra credit parts are often open-ended. How much they should count and how they should be graded is up to individual instructors. Some way has to be found to get students to take these parts seriously. If students just take a stab at it, most likely what they turn in will be garbage. With hints given in class the students do better.

While there are no recipes for originality, this chapter should help you create workable projects.

Part II

GUIDE TO THE PROJECTS

Here we mention a few things you should know about our projects before selecting one for classroom use. You should also read the chapter "Logistics: Ideas for Using Projects Successfully" before actually assigning a project to students.

There is considerable variety to the projects in this book. They range from a few precalculus projects, which can be assigned at the beginning of calculus, through single variable and on to multivariate calculus topics. Their pedagogical spirit stretches from "minimalist" statements to carefully guided road maps, and their context ranges from pure mathematics to "pseudo-applied" projects and those with story-lines. Some of our projects have been inspired by journals and books, although we find more and more that project ideas simply pop to mind now that we are seeking them.

We offer a guide rating the difficulty of each project based on the amount of instructor assistance students have needed. Of course, these ratings are more relative than absolute, and you will need to adjust for your students and situation. Each project has been rated from 1 through 4, and the ratings are indicated in the "List of Projects." These ratings have the following meaning:

1. Most students require little help.
2. Most students require moderate help.
3. Most students require considerable help.
4. Most students require a lot of help.

We have arranged the projects in roughly the order calculus is typically taught, although this is sometimes ill-defined since the nature of projects is that they often require and teach more than one topic. Each project has a descriptive name, as shown in the "List of Projects."

The projects are ready to be photocopied directly for assignment to students, although after each project there is information for you, the instructor, which you will wish to exclude when copying. This information begins with two sets of keywords, the first alerting you to the topics that are required knowledge for working on the particular project, and the second set giving the major topics that the project teaches. Refer to the Index to find those projects that require or teach particular topics.

After these two sets of key words for each project you will also find our notes to instructors about the project. These notes provide information you should know before assigning the project. Most of these notes are based on actual classroom experience with the project, and thus give hints and advice about how the students may react to the mathematics, how you can best assist them in their discoveries, and what traps you can help them avoid. If a project has never been assigned this is mentioned in the notes.

The instructor notes for projects include several other things. Most of our projects are intended as two-week assignments, but some are only one-week projects. The one-week assignments are identified in the notes, and are marked by a (∗) in the "List of Projects." While most of our projects were designed for a calculus course for scientists, mathematicians, and engineers, some are for a course for students from biology, business, and social sciences, and these are also specially mentioned.

Some of our projects were created in pairs and trios for simultaneous assignment to portions of a class. They have complementary mathematical content that may allow a special lecture to be built around the results of the entire group of projects. Each such group of projects is indicated by a brace ({) enclosing their project numbers in the "List of Projects," and a common set of notes to instructors discusses the group of projects as a whole. This common set of notes begins after all the projects in the group.

Finally, some projects have extra credit challenges at the end, which may be quite open-ended. In the spirit of exploration through projects, we may not have a particular solution in mind. You will have to decide how to handle these extra credit suggestions, since they might lead you and your students into unexpected and unexplored territory.

List of Projects[1][2]

---

[1] The numbers in the Rating column give the relative difficulty of projects, from easiest (1) to hardest (4). See the "Guide to the Projects" for details.

[2] An asterisk (*) denotes a one-week, as opposed to two-week, project. A brace ({) groups projects with complementary mathematical content. See the "Guide to the Projects" for details.

The Volumes of a Regular Tetrahedron and a Regular Octahedron

The volume of either a regular tetrahedron or a regular octahedron is difficult to calculate because the angles between faces are not right angles. The volume of each figure will be calculated in two different ways, using trigonometry and insight. The latter method requires less computation. We assume that the formula for the volume of a pyramid is known.

I. The regular tetrahedron

    a. Calculate its volume by trigonometry.

    b. This figure can be embedded in a cube by letting each edge of the tetrahedron be a diagonal of a face of the cube. (Two different embeddings are possible.) Left over are several tetrahedra whose volumes are easy to compute since they have some right angles. Now again calculate the volume of a regular tetrahedron, without using trigonometry or calculus.

II. The regular octahedron.

    a. Calculate its volume by trigonometry.

    b. This figure can be embedded in a cube by letting each vertex of the octahedron be at the center of a face of the cube. This octahedron is the intersection of the two tetrahedra of part (Ib). Now again calculate the volume of a regular octahedron without using trigonometry or calculus.

Comments

Requires: volume of a pyramid, trigonometry

Teaches: spatial visualization

    This may be a good one-week project. As stated, the project requires no calculus. However, it may be modified to do so by having students first derive the assumed volume formula, either directly for pyramids or, more broadly, for generalized cones; these are both standard problems stated in a number of calculus texts.

Evaluating Polynomials

Computers use procedures, called algorithms, to evaluate polynomials. We are going to study the efficiency of various algorithms for evaluating polynomials. We will do this by counting the number of basic operations needed; since multiplication takes much more time to perform on a computer, we will count *only* multiplications.

a. How many multiplications does it take to evaluate the one-variable polynomial,

$$a_0 + a_1 x + a_2 x^2 + \cdots + a_n x^n = \sum_{i=0}^{n} a_i x^i,$$

   when the operations are performed as indicated? (Remember that powers are repeated multiplication and must be counted as such.) Write this number of multiplications as a function of n.

b. Use mathematical induction to prove that your answer is correct.

c. Find another way to evaluate this polynomial by doing the operations in a different order so that fewer multiplications are required. Try to think of ways to intermix addition and multiplication. Write the number of multiplications as a new function of n. The cheapest involves only n multiplications. Find such an algorithm.

d. How many multiplications does it take to evaluate the two-variable polynomial,

$$\sum_{i=0}^{n} \sum_{j=0}^{n} a_{ij} x^i y^j,$$

   when the operations are performed as indicated? Write this number of multiplications as another function of n.

e. Use mathematical induction to prove that your answer is correct.

f. Find another way to evaluate this two-variable polynomial by doing the operations in a different order so that fewer multiplications are required. Write down the associated function of n. Is this the cheapest? If not, find the cheapest.

Extra credit: Redo parts (d), (e), and (f) for the polynomial $\sum_{i=0}^{m} \sum_{j=0}^{n} a_{ij} x^i y^j$ with $m \neq n$.

Comments

Requires: mathematical induction

Teaches: mathematical induction, counting arguments

This project has never been assigned.

Here is a noncalculus project that nevertheless could be assigned as the first project in the first calculus course. It may also be appropriate when studying infinite series. The project will review algebra as well as teach the student about making up formulae to describe the efficiency of an algorithm. Also the student will have to look up induction and learn how to write induction proofs.

It is difficult to prove that the lower bounds of (c) and (f) are the best. Hence this was not called for.

Composing Functions

Consider the two functions:

$$f(x) = 1 - x \quad \text{and} \quad g(x) = \frac{1}{x}$$

We can compose them in two ways:

$$f(g(x)) \quad \text{and} \quad g(f(x)).$$

We can go further and compose these two new functions with themselves, and also with the old ones, in a number of ways. Keep composing these functions with new ones as they are generated and figure out simplified formulae for them in terms of the variable x. (Don't forget to compose functions with themselves, like $f(f(x))$.) You might think that more and more new functions will be generated. Surprisingly only a finite number of new ones get generated by composition, even though there may be many different ways of composing f and g to get the same function. Remember that two very different looking formulae may represent the same function.

   a. How many distinct functions are there, including f and g themselves?
   b. List them.
   c. How is each one composed from f and g?
   d. How do you know that these are all there are?
   e. For what real numbers are all these functions simultaneously defined?

Comments

Requires: composition of functions

Teaches: composition of functions, equality of functions, finite group theory, non-commutativity

   Students learned a lot from this project about functions and they liked it. This project has been used in both calculus for engineers and calculus for biologists.
   Once an instructor appreciates the underlying group theory, it is possible to create any number of similar projects. The functions above are examples of bilinear transformations,

$$f(x) = \frac{ax + b}{cx + d};$$

these can be represented by invertible matrices: $A = \begin{bmatrix} a & b \\ c & d \end{bmatrix}$, where $|A| \neq 0$. Composition of bilinear transformations is represented by multiplication of matrices: $f_A \circ f_B = f_{AB}$. Invertible matrices make up the classical general linear group $GL_n(\mathbb{R})$. But the bilinear transformations are less than this since there is an equivalence relation: $f_A = f_B$ if and only if $A = cB$ for some $c \neq 0$. We need only consider matrices with determinant ± 1, and identify those that are negatives of each other. This amounts to looking at a homomorphic image of a subgroup of $GL_2(\mathbb{R})$:

$$\frac{\{A \in GL_2(\mathbb{R}) : |A| = \pm 1\}}{\left\{ \begin{bmatrix} 1 & 0 \\ 0 & 1 \end{bmatrix}, \begin{bmatrix} -1 & 0 \\ 0 & -1 \end{bmatrix} \right\}}$$

That is, this quotient group is isomorphic to the group of bilinear transformations on $\mathbb{R}$. It is an extension of the projective special linear group by the cyclic group of order 2. The projective special linear group itself, $PSL_2(\mathbb{R})$, is isomorphic to the group of bilinear transformations with positive determinant.

Another example of a group of bilinear transformations coming out of this analysis that would be suitable as a project is given by the generators: $f(x) = \frac{x+1}{-x+1}$ and $g(x) = -x$. These were created by thinking of the corresponding matrices as linear transformations of the plane into itself. We wanted something of order four together with a noncommuting involution. But a rotation matrix of order four would yield a bilinear transformation of order two because of the halving of the order as explained before. Hence it was necessary to create a rotation of order eight; its rationalization as a bilinear transformation is f and it has order four.

A third example is given by the generators $f(x) = 2x$ and $g(x) = \frac{1}{x}$. We leave it to the interested instructor to figure out the groups generated in all these examples.

Continuity of Compositions

In this project you will learn about the precise definition of a limit, and explore some basic features of limits in a mathematically rigorous fashion, rather than in the intuitive, informal fashion we have been using so far. It took 200 years, from the invention of calculus in the 17th century to the 19th century, for mathematicians to place the notion of a limit on a firm footing, so do not be surprised if it takes a lot of intense concentration to understand it properly. By the time you are finished, though, you will have really mastered the intricacy and intrigue of the idea of a limit. You will feel that you have really accomplished something!

First read about the precise definition of a limit, and do various exercises to develop your expertise and confidence. In what follows, do not assume or use any informal or intuitive arguments about limits. Use only the precise definitions of limits in everything you do. Do not even make use of any theorems about limits. Just work from scratch using the definitions. Draw graphs to help yourself see what is happening.

a. Show that the function $2x - 3$ is continuous at every real number.

b. Show that the function x^2 is continuous at $x = 1$; at $x = 3$; at any real number x.

c. Show that if g is continuous at a, and if f is continuous at $g(a)$, then the composite function, $h = f \circ g$, is continuous at a.

d. Explain what parts (a), (b), and (c) will together imply about the functions $(2x - 3)^2$ and $2x^2 - 3$.

Limit of a Reciprocal

In this project you will learn about the precise definition of a limit, and explore some basic features of limits in a mathematically rigorous fashion, rather than in the intuitive, informal fashion we have been using so far. It took 200 years, from the invention of calculus in the 17th century to the 19th century, for mathematicians to place the notion of a limit on on a firm footing, so do not be surprised if it takes a lot of intense concentration to understand it properly. By the time you are finished, though, you will have really mastered the intricacy and intrigue of the idea of a limit. You will feel that you have really accomplished something!

First read about the precise definition of a limit, and do various exercises to develop your expertise and confidence. In what follows do not assume or use any informal or intuitive arguments about limits. Use only the precise definitions of limits in everything you do. Do not even make use of any theorems about limits. Just work from scratch using the definitions. Draw graphs to help yourself see what is happening.

a. Show that the function $2x - 3$ is continuous at every real number.

b. Show that $\lim_{x \to 2} \frac{1}{x} = \frac{1}{2}$.

c. Show that if $\lim_{x \to a} f(x)$ exists and is not zero, then $\lim_{x \to a} \left(\frac{1}{f(x)} \right)$ exists and

$$\lim_{x \to a} \left(\frac{1}{f(x)} \right) = \frac{1}{\lim_{x \to a} f(x)}.$$

Hint: First give $\lim_{x \to a} f(x)$ a simple name.

d. Combine your results to show that the function $\frac{1}{2x-3}$ is continuous at every real number except $\frac{3}{2}$.

Hint: Start by naming the function in part (a).

Algebra of Infinite Limits

In this project you will learn about the precise definition of a limit, and explore some basic features of limits in a mathematically rigorous fashion, rather than in the intuitive, informal fashion we have been using so far. It took 200 years, from the invention of calculus in the 17th century to the 19th century, for mathematicians to place the notion of a limit on a firm footing, so do not be surprised if it takes a lot of intense concentration to understand it properly. By the time you are finished, though, you will have really mastered the intricacy and intrigue of the idea of a limit. You will feel that you have really accomplished something!

First read about the precise definition of a limit, and do various exercises to develop your expertise and confidence. In what follows do not assume or use any informal or intuitive arguments about limits. Use only the precise definitions of limits in everything you do. Do not even make use of any theorems about limits. Just work from scratch using the definitions. Draw graphs to help yourself see what is happening.

a. Develop and explain a precise definition of $\lim_{x \to a} f(x) = \infty$.

   In what follows, use your definition.

b. Suppose that $\lim_{x \to a} f(x) = \infty$ and $\lim_{x \to a} g(x) = \infty$. Make a conjecture about $\lim_{x \to a}(f(x) + g(x))$. Prove your conjecture.

c. Suppose that $\lim_{x \to a} f(x) = A$, where A is a real number, and $\lim_{x \to a} g(x) = \infty$. Make a conjecture about $\lim_{x \to a}(f(x) + g(x))$. Prove your conjecture.

d. Suppose that $\lim_{x \to a} f(x) = A$, where A is a positive real number, and that $\lim_{x \to a} g(x) = \infty$. Make a conjecture about $\lim_{x \to a}(f(x) \cdot g(x))$. Prove your conjecture.

Comments on *Continuity of Compositions, Limit of a Reciprocal,* and *Algebra of Infinite Limits*

Requires: limits

Teaches: precise definition of limit, continuity, composition of functions, infinite limits

These three projects focus on the precise definition of a limit. The instructor can give a lecture that helps students working on them to get over initial hurdles. These projects are quite a challenge for most students and they may need substantial assistance.

The Case of the Smudged Map

You are the owner of a small surveying firm, struggling to make a name for yourself in the community, and you are facing a crisis. In the settlement of an estate, the judge has decreed that a wedge-shaped piece of property (shaped like a piece of pie) be divided in two parts by a straight fence starting at the corner of the wedge. This dividing fence must be built such that if an east-to-west fence and a north-to-south fence are constructed from any point on it, the resulting regions beginning at the corner of the wedge have equal areas. Your task is to design the fence and set out stakes for its builders.

The problem is that you left your truck window open and rain has smudged the ink on the map. You don't have time to drive to the site and make measurements; the job must be done right away. All you can see from the map is that the two edges of the wedge-shaped property are straight and proceed in directions somewhere between north and east from a common point.

Design the fence. If you do a good job, this might be the turning point in your career.

Comments

Requires: lines, area

Teaches: arbitrary constants

This is a somewhat easier version of a project assigned in an accelerated calculus course. This project could be used in a precalculus course and should be a one-week project.

Limits of Functions that Convert Addition to Multiplication

Suppose that a certain city has an initial population of 10,000 and that the population increases by 10% every year. After one year, the population will be

$$10{,}000 + (.10)(10{,}000) = 11{,}000.$$

After two years, the population will be

$$11{,}000 + (.10)(11{,}000) = 12{,}100.$$

a. What will the population be after 3 years?

Let $P(x)$ be a function that gives the population of this city after x years.

b. Show that

$$P(2) = \frac{1}{10{,}000} P(1)P(1) \quad \text{and} \quad P(3) = \frac{1}{10{,}000} P(2)P(1).$$

In general we have

$$P(x+y) = \frac{1}{10{,}000} P(x)P(y),$$

although you do not have to show this. Notice that the constant $1/10{,}000$ in front of this equation appears since the initial population is given as 10,000. In some sense this number is artificial, since the initial population could be any number. If we normalize the initial population to be 1, then $P(x+y) = P(x)P(y)$. This is the key property of this function. Note how the function P can be computed for future years $x+y$ by multiplying the values of the population for the year x and the year y. Such a function can be described as converting addition, $P(x+y)$, to multiplication, $P(x)P(y)$.

Suppose $g : \mathbb{R} \to \mathbb{R}$ is a function that satisfies $g(x+y) = g(x)g(y)$ and $g(1) = b$ where $b \neq 0$. Here $\mathbb{R}$ denotes the set of real numbers.

c. Show that $g(x) \neq 0$ for all $x \in \mathbb{R}$.

d. Suppose that $\lim_{x \to 0} g(x) = g(0)$. By the definition of limit, given $\epsilon > 0$, there is some $\delta > 0$ such that $|g(x) - g(0)| < \epsilon$ whenever $0 < |x - 0| < \delta$. Use the definition of limit to show that $\lim_{x \to a} g(x) = g(a)$ for any a. What does your result say about the continuity of g?

Hint: Try to translate the ϵ–δ proof from the point $x = a$ to the point $x = 0$.

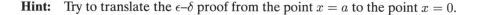

The Form of Functions that Convert Addition to Multiplication

Suppose that a certain city has an initial population of 10,000 and that the population increases by 10% every year. After one year, the population will be

$$10{,}000 + (.10)(10{,}000) = 11{,}000.$$

After two years, the population will be

$$11{,}000 + (.10)(11{,}000) = 12{,}100.$$

a. What will the population be after 3 years?

   Let $P(x)$ be a function that gives the population of this city after x years.

b. Show that

$$P(2) = \frac{1}{10{,}000}P(1)P(1) \quad \text{and} \quad P(3) = \frac{1}{10{,}000}P(2)P(1).$$

   In general we have that

$$P(x + y) = \frac{1}{10{,}000}P(x)P(y),$$

although you do not have to show this. Notice that the constant 1/10,000 in front of this equation appears since the initial population is given as 10,000. In some sense this number is artificial, since the initial population could be any number. If we normalize the initial population to be 1, then $P(x + y) = P(x)P(y)$. This is the key property of this function. Note how the function P can be computed for future years $x + y$ by multiplying the values of the population for the year x and the year y. Such a function can be described as converting addition, $P(x + y)$, to multiplication, $P(x)P(y)$.

   Suppose $g : \mathbb{R} \to \mathbb{R}$ is a function that satisfies $g(x + y) = g(x)g(y)$ and $g(1) = b$ where $b \neq 0$. Here $\mathbb{R}$ denotes the set of real numbers.

c. Show that $g(x) \neq 0$ for all $x \in \mathbb{R}$.

d. Show that $g(s) = b^s$ for all rational numbers s, given that $g(1) - b$ where $b \neq 0$.

Hint: It may help to argue as follows:

1. Show that $g(0) = 1$.

2. Let n be a positive integer. Show that:

$$g(n) = b^n, \quad g(-n) = b^{-n}, \quad \text{and} \quad g(1/n) = b^{1/n}.$$

3. For integers n and m where $m \neq 0$, show that $g(n/m) = b^{n/m}$.

Approximating Irrational Numbers

The goal of this project is to show that every irrational number can be approximated by a sequence of rational numbers. You will need the following definitions. A *sequence* is a function with domain the set of positive integers; often it is written as $x_1, x_2, x_3, \ldots$. For example, the formula $x_n = 1 + \frac{1}{n}$ defines a sequence with $x_1 = 1 + 1 = 2$, $x_2 = 1 + \frac{1}{2} = \frac{3}{2}$, $x_3 = 1 + \frac{1}{3} = \frac{4}{3}$, and so on. We can define limits of sequences in a manner similar to that used to define limits of functions. For a sequence $\{x_n\}_{n=1}^{\infty}$, we say that $\lim_{n\to\infty} x_n = L$ if, given $\epsilon > 0$, there is some positive integer N such that $|x_n - L| < \epsilon$ whenever $n \geq N$. To illustrate how to use this definition, we shall prove the following result for you.

Claim. $\lim_{n\to\infty} \left(1 + \frac{1}{n}\right) = 1$.

Proof.   Suppose that $\epsilon > 0$. We must find a positive integer N so that $\left|\left(1 + \frac{1}{n}\right) - 1\right| < \epsilon$ whenever $n \geq N$. Choose an integer N such that $N > \frac{1}{\epsilon}$. Then $\frac{1}{N} < \epsilon$. If $n \geq N$, then $\frac{1}{n} \leq \frac{1}{N}$ so that $\left|\left(1 + \frac{1}{n}\right) - 1\right| = \left|\frac{1}{n}\right| = \frac{1}{n} \leq \frac{1}{N} < \epsilon$. Thus, $\lim_{n\to\infty}(1 + \frac{1}{n}) = 1$.

Your project is to use the definition for the limit of a sequence to prove that given an irrational number α, there is a sequence of rational numbers $\{x_n\}_{n=1}^{\infty}$ such that $\lim_{n\to\infty} x_n = \alpha$.

Hint: Suppose the decimal expansion of a positive irrational number α is $\alpha = n_0.n_1n_2n_3\ldots$, where n_0 is the largest integer less than α, and $n_1, n_2, n_3, \ldots$ are the digits in a decimal expansion for α. For example $\pi = 3.141\ldots$, $n_0 = 3, n_1 = 1, n_2 = 4$, $n_3 = 1$, and so on.

You may assume that given $\epsilon > 0$, there is some integer N with $\frac{1}{\epsilon} < 10^N$. Why is it enough to consider only positive irrational numbers?

Comments on *Limits of Functions that Convert Addition to Multiplication, The Form of Functions that Convert Addition to Multiplication,* **and** *Approximating Irrational Numbers*

Requires: precise definition of limit, decimal representation of real numbers

Teaches: precise definition of limit

These three projects were assigned at the same time. Some class time was spent putting the three results together to show that a nonzero function that converts addition to multiplication and is continuous at the origin must be an exponential function.

Instructors should be aware of one difficulty with the project *Limits of functions that convert addition to multiplication* (#8). Students must fix an ϵ and then determine a new ϵ' that depends on ϵ and $g(a)$.

Continuity and a Very Curious Function

A real number x is called a *rational number* if there are integers m and n such that $x = \frac{m}{n}$. In the following, assume that $\frac{m}{n}$ is a fraction reduced to lowest terms. A fraction $\frac{m}{n}$ is in lowest terms when $n > 0$ and m and n have no common factors. A real number that is not a rational number is called an *irrational number*. Define a function f on the interval $(0, 1)$ by

$$f(x) = \begin{cases} 0 & \text{if } x \text{ is an irrational number} \\ \frac{1}{n} & \text{if } x \text{ is a fraction, } \frac{m}{n}. \end{cases}$$

a. Determine where f is continuous and where f is discontinuous.

b. Can you define $f(0)$ so that f will be continuous from the right at 0?

Use the precise definition of limit to convince the reader of all your answers to parts (a) and (b).

Hint: You may use without proof the fact that every interval of real numbers contains both rational and irrational numbers.

Comments

Requires: limits, continuity

Teaches: precise definition of limit, continuity

This project has only been assigned as extra credit and is difficult. You may wish to give the following hint for showing continuity at the irrational numbers: how many rational numbers (in lowest terms) between 0 and 1 are there with denominator exactly N? How many are there with denominator at most N?

A Continuous Additive Function

In this project you will learn about additive functions and discover that continuous additive functions have a very special form.

Definition.   Let f be a function whose domain is the set of all real numbers. Then f is *additive* if $f(x + y) = f(x) + f(y)$ for all real numbers x and y .

a. Give an example of an additive function and show that it is additive. Give an example of a function that is not additive and show that it is not additive.

b. Suppose that f is an additive function and m is any real number. Define a new function g by the formula $g(x) = f(x) - mx$. Show that g is an additive function.

c. Let $m = f(1)$ and show that the function g, defined in part (b) above, has the property that $g(x + 1) = g(x)$ for all x .

Definition.   A function f is *bounded* on a closed interval $[a, b]$ if there is a real number B such that $|f(x)| \leq B$ for all x in $[a, b]$. A function f is *bounded* if there is a real number B such that $|f(x)| \leq B$ for all x.

d. Suppose that f is an additive function that is bounded on $[0, 1]$ and g is defined by $g(x) = f(x) - mx$ where $m = f(1)$. Show that g is bounded.

e. Let g be a bounded additive function; that is, there is a real number B such that $|g(x)| \leq B$ for all x. If there is a real number a such that $g(a)$ is not zero, show that there is a real number t such that $|g(t)| > B$, a contradiction. What can you conclude about g? Explain.

   Hint: What is $g(2a)$, what is $g(3a)$?

   At the beginning of the project, you were told that you would discover that continuous additive functions have a very special form. As you might suspect, the preceding parts of this project are designed to help you in that discovery. You will need to use some facts about continuous functions.

f. Show that if f is additive and continuous, there is a real number m such that $f(x) = mx$ for all x .

A Continuous Affine Function

In this project you will learn about affine functions and discover that continuous affine functions have a very special form.

Definition. Let f be a function whose domain is the set of all real numbers. Then f is *affine* if $f(x - y + z) = f(x) - f(y) + f(z)$ for all real numbers x, y, and z.

a. Give an example of an affine function and show that it is affine. Give an example of a function that is not affine and show that it is not affine.

Definition. Let g be a function whose domain is the set of all real numbers. Then g is *additive* if $g(x + y) = y(x) + g(y)$ for all real numbers x and y.

b. Suppose that f is an affine function and m is any real number. Define a new function g by the formula $g(x) = f(x) - (mx + f(0))$. Show that g is an additive function.

c. Let $m = f(1) - f(0)$ and show that the function g defined in part (b) above has the property that $g(x + 1) = g(x)$ tor all x .

Definition. A function f is *bounded* on a closed interval $[a, b]$ if there is a real number B such that $|f(x)| \leq B$ for all x in $[a, b]$. A function f is *bounded* if there is a real number B such that $|f(x)| \leq B$ for all x .

d. Suppose that f is an affine function that is bounded on $[0, 1]$ and g is defined by $g(x) = f(x) - (mx + b)$ where $m = f(1) - f(0)$ and $b = f(0)$. Show that y is bounded.

e. Let g be a bounded additive function; that is, there is a real number B such that $|g(x)| \leq B$ for all x . If there is a real number a such that $g(a)$ is not zero, show that there is a real number t such that $|g(t)| > B$, a contradiction. What can you conclude about g? Explain.

Hint: What is $g(2a)$, what is $g(3a)$?

At the beginning of the project, you were told that you would discover that continuous affine functions have a very special form. As you might suspect, the preceding parts of this project are designed to help you in that discovery. You will need to use some facts about continuous functions.

f. Show that if f is affine and continuous, there are real numbers m and b such that $f(x) = mx + b$ for all x.

Comments on *A Continuous Additive Function* and *A Continuous Affine Function*

Requires: functions, functional notation, maximum/minimum value theorem, absolute value, inequalities, bounded function

Teaches: continuous functions, functional notation

These are both difficult projects. They were assigned in a typical first semester calculus class and the students did moderately well with a *lot* of instructor help. For most students, the hard part was simply understanding the language. They have little experience with collections of words like "for all" and "there is." These projects should be used with above average classes. Similar results hold for functions on $\mathbb{R}^n$, so the project can be adapted to more advanced courses.

An Increasing Function and the Most Important Limit in Calculus

Suppose that you deposit A dollars in a bank account at a fixed interest rate r (for instance, $r = 0.09$ corresponds to 9% interest). At the end of one year, you have $A(1+r)$ dollars. If the bank compounds interest quarterly, however, you would have $A\left(1+\frac{r}{4}\right)^4$ dollars. With daily interest, after one year you would have $A\left(1+\frac{r}{365}\right)^{365}$ dollars. Today, many banks claim to compound interest continuously, so one is led to consider the limit of $A\left(1+\frac{r}{y}\right)^y$ as y approaches infinity.

To simplify things, we first set $x = \frac{y}{r}$ and then rewrite this expression as

$$A\left(1+\frac{r}{y}\right)^y = A\left\{\left(1+\frac{1}{x}\right)^x\right\}^r.$$

This leads us to analyze the limit $\lim_{x\to\infty}\left(1+\frac{1}{x}\right)^x$. Your project is concerned with this limit. This limit is important not only in banking, but also in mathematics and science; for instance, it arises in the study of radioactive decay, population growth, and electronics.

We will make a simplification by considering only positive integral values for x. So, let n be a positive integer and consider $\left(1+\frac{1}{n}\right)^n$. You will prove that $\left(1+\frac{1}{n}\right)^n$ is an increasing function of n.

To begin your work, expand both $\left(1+\frac{1}{n}\right)^n$ and $\left(1+\frac{1}{n+1}\right)^{n+1}$ using the binomial theorem. Show that the latter expression is larger than the former by comparing the two expansions term by term.

Hint: Note that $\dfrac{n-k}{n} = 1 - \dfrac{k}{n}$.

A Bounded Function and the Most Important Limit in Calculus

Suppose that you deposit A dollars in a bank account at a fixed interest rate r (for instance, $r = 0.09$ corresponds to 9% interest). At the end of one year, you have $A(1 + r)$ dollars. If the bank compounds interest quarterly, however, you would have $A\left(1 + \frac{r}{4}\right)^4$ dollars. With daily interest, after one year you would have $A\left(1 + \frac{r}{365}\right)^{365}$ dollars. Today, many banks claim to compound interest continuously, so one is led to consider the limit of $A\left(1 + \frac{r}{y}\right)^y$ as y approaches infinity.

To simplify things, we first set $x = \frac{y}{r}$ and then rewrite this expression as

$$A\left(1 + \frac{r}{y}\right)^y = A\left\{\left(1 + \frac{1}{x}\right)^x\right\}^r .$$

This leads us to analyze the limit $\lim_{x \to \infty}\left(1 + \frac{1}{x}\right)^x$. Your project is concerned with this limit. This limit is important not only in banking, but also in mathematics and science; for instance, it arises in the study of radioactive decay, population growth, and electronics.

We will make a simplification by considering only positive integral values for x. So, let n be a positive integer and consider $\left(1 + \frac{1}{n}\right)^n$. You will prove that, as a function of n, $\left(1 + \frac{1}{n}\right)^n$ is bounded from above.

Definition.   A function f is *bounded from above* if there is a real number B such that $f(x) \le B$ for all x in the domain of f. Such a number B is called an *upper bound* of f.

To begin your proof that $\left(1 + \frac{1}{n}\right)^n$ is bounded from above, first prove that

$$1 + r + r^2 + \cdots + r^n = \frac{1 - r^{n+1}}{1 - r}$$

for any r, $0 \le r < 1$. Next, use the binomial theorem to show that

$$\left(1 + \frac{1}{n}\right)^n \le 1 + \left\{1 + \frac{1}{2} + \left(\frac{1}{2}\right)^2 + \cdots + \left(\frac{1}{2}\right)^{n-1}\right\}.$$

Explain why this proves that $\left(1 + \frac{1}{n}\right)^n$ is bounded from above. What number did you find for an upper bound?

The Most Important Limit in Calculus Exists!

Suppose that you deposit A dollars in a bank account at a fixed interest rate r (for instance, $r = 0.09$ corresponds to 9% interest). At the end of one year, you have $A(1 + r)$ dollars. If the bank compounds interest quarterly, however, you would have $A \left(1 + \frac{r}{4}\right)^4$ dollars. With daily interest, after one year you would have $A \left(1 + \frac{r}{365}\right)^{365}$ dollars. Today, many banks claim to compound interest continuously, so one is led to consider the limit of $A \left(1 + \frac{r}{y}\right)^y$ as y approaches infinity.

To simplify things, we first set $x = \frac{y}{r}$ and then rewrite this expression as

$$A \left(1 + \frac{r}{y}\right)^y = A \left\{ \left(1 + \frac{1}{x}\right)^x \right\}^r.$$

This leads us to analyze the limit $\lim_{x \to \infty} \left(1 + \frac{1}{x}\right)^x$. Your project is concerned with this limit. This limit is important not only in banking, but also in mathematics and science; for instance, it arises in the study of radioactive decay, population growth, and electronics.

Definition. A function f is *bounded from above* if there is a real number B such that $f(x) \leq B$ for all x in the domain of f. Such a number B is called an *upper bound* of f.

For $x \geq 1$, $\left(1 + \frac{1}{x}\right)^x$ is bounded from above and increasing. You do not need to show this. In your project you will prove that any function that is bounded from above and increasing has a limit at infinity. In particular, $\lim_{x \to \infty} \left(1 + \frac{1}{x}\right)^x$ exists.

If a function has an upper bound, then it has many. There is one upper bound that is of particular importance.

Definition. A number L is called the *least upper bound* of the function f if L is an upper bound of f, and L is less than every other upper bound of f.

You will need to use a fundamental property of the real numbers known as the *least upper bound property*.

Least upper bound property. *Every function that is bounded from above has a least upper bound.*

The following lemma and the proof we provide for it will give you a better understanding of least upper bounds. You may use this lemma in your work.

Lemma. *If L is the least upper bound of a function f, and $M < L$, then there is an x in the domain of f so that $M < f(x)$.*

Proof.   By the assumptions that $M < L$ and that L is the least upper bound of f, we know that M is not an upper bound of f. Since M is not an upper bound of f, the inequality $f(x) \leq M$ does not hold for all x in the domain of f. Thus there must be some x in the domain of f so that $f(x) > M$. This completes the proof of the lemma.

Now, your task in this project is to prove the following theorem yourself, using the rigorous definition of limits.

Theorem.   *Suppose f is a function on (a, ∞) that is both bounded from above and increasing. Let L be the least upper bound of f on (a, ∞). Then*

$$\lim_{x \to \infty} f(x) = L.$$

Comments on *An Increasing Function and the Most Important Limit in Calculus, A Bounded Function and the Most Important Limit in Calculus,* and *The Most Important Limit in Calculus Exists!*

Requires: limits at infinity, binomial theorem

Teaches: precise definition of limits, index notation, factorial, binomial theorem, least upper bound property

These three projects were assigned at the same time. After the completion of the projects, some class time was spent putting the three results together to discuss the number e. This was done about two months before we normally cover e in the course.

The first two of the projects required students to learn to handle symbolic notation. This is hard for students, and is a good assignment to give them early in calculus. The last project involved a proof using the definition of limits at infinity. The definition used for the least upper bound is nonstandard. We feel that wording the problem this way simplifies things for the students.

Finding the Area of a Circle

Everyone knows that the area of a circle of radius r is πr^2. Or, at least, everyone thinks they know that result is correct because everyone has been told so by lots of teachers. But if you happened to live before 300 B.C., you would not know the formula for the area of a circle.

It was Archimedes who derived the area of the circle, around 250 B.C. Here is a method similar to the one he used. Let n be a positive integer and cut the circle into n equal sectors. In each sector there is an isosceles triangle formed by the center of the circle and the points where the straight edges of the sector cut the circumference. He found the area of each triangle and then approximated the area of the circle by the sum of the areas of the triangles. Finally, he let the number of triangles (n) increase without bound to find the area of the circle.

Your goal is to repeat this procedure and prove that the formula for the area of a circle is correct. It may help you to figure out the central angle (the angle at the center of the circle) of each sector. What happens to this angle as the number of triangles increases without bound?

Since your answer agrees with your expectations, you probably think that you have found the area of the circle. But, you must be sure that the process above does not "miss any of the area." To check this, you will estimate how far off your approximation of the area of the circle is and show that this error approaches 0 as the number of triangles increases without bound. Divide the circle into n sectors and form the inscribed triangles as before. At the base of each isosceles triangle, form the smallest rectangle that contains the area of the sector that is outside of the triangle. What is the area of this rectangle? Use this to bound the error in approximating the area of the sector by the area of the triangle. Find a bound for the total error in approximating the area of the circle by the sum of the areas of the triangles. Show that this error approaches 0 as the number of triangles increases without bound, which proves that your formula for the area of the circle is correct.

Extra credit: Can you think of other simple geometric shapes that could be used instead of triangles? Would they be easier or harder to use? Why? Where would you run into difficulties with the computations you did above?

Comments

Requires: trigonometry, limits

Teaches: trigonometric identities, error estimation

In order to solve this problem, the students need to know the limits of $\frac{\sin x}{x}$ and $\frac{\cos x - 1}{x}$ as x approaches 0. The problem makes them face the difference between being told what is true and having to prove something is true on their own. The main difficulty students had with this problem was their inability to read and comprehend.

The Derivative of the Three-halves Power

The power rule gives the derivative of the function, $f(x) = x^{3/2}$, as $\frac{3}{2}x^{1/2}$, when $x > 0$. The object of this project is to verify this special case of the power rule using the definition of the derivative given in terms of a limit. To do this completely, you will need some of the properties of limits. You may accept and use these without proving them. To tackle this project, it is recommended that you first derive the derivatives in the simpler cases: $f(x) = x^{1/2}$ and $f(x) = x^3$.

Comments

Requires: limits, derivatives

Teaches: limits, definition of derivative

This was assigned in a course for budding biologists and social scientists, and would not be suitable in calculus for engineers, physical scientists, and mathematicians.

Textbooks for calculus in the biological and social sciences do not always give an unambiguous and clear presentation of the notion of limit for students to understand and develop proofs as we expect. Before assigning, check carefully how well this project will fit your text. Our experience with the students' use of the properties of limits was negative. But this was not entirely the students' fault. The section on limits in the text we were using was a mishmash of example, definition, theorem, and motivation without a clear indication of what was what.

Preparing for the 1990 United States Census

The government census bureau is preparing to take the census of the United States in 1990, and you have been hired to help them plan for the counting of the American population. You job is to make a projection of the population so that the bureau will know how many census takers to hire.

Your director, Dr. John Knowitall, wants you to proceed as follows. He wants you to use the actual population data from two consecutive polls, along with the actual rates of population growth at the times of the two polls. He wants you to find a quadratic function that models the population as a function of time and agrees with the actual populations and the rates of growth of the two polls.

a. Your first task in this project is to show Dr. Knowitall that in general it is impossible to find such a function. To do this, let t_0 be the year of the first poll, and let t_1 be the year of the second poll. Also, let p_0 and p_1 denote the population values from the polls and let r_0 and r_1 denote the rates of growth from the two polls. To simplify your computations, write your polynomial in the variable $t - t_0$. (If you do not think this really does simplify the arithmetic, repeat part (a) using a polynomial in the variable t and see what kind of computations are needed.)

Being the resourceful person that you are, you realize that the difficulty was caused by the fact that the quadratic function was a polynomial of too low a degree to allow you to include all the information that Dr. Knowitall wants.

b. Show that by increasing the degree of the polynomial to three, it is always possible to fit the data to a new polynomial (that is, to find a polynomial that agrees with the populations and the rates of growth of any two polls).

c. After completing parts (a) and (b), use the actual census data below along with your results from part (b) to find population functions for each pair of consecutive poll data. Use this to project the population of the United States for 1990, as follows. First use your polynomial modelling function for each two consecutive polls to project the population of the succeeding census and compare the result to the actual census taken. Use all this information to make your projection for 1990. How confident are you of your projection? Discuss why.

Year	Population	Rate of growth of population (people per year)
1950	151,325,798	2,357,930
1960	179,323,175	2,598,812
1970	203,302,031	2,360,967
1980	226,542,518	2,324,049

Comments

Requires: derivatives, systems of linear equations

Teaches: arbitrary constants

We assigned a second version of this project along with this one, using census data for the state of New Mexico. The quality of the fitting by splines depends very much on the initial data and can vary considerably from state to state. In part (c), students are asked to pass judgement on the quality of the fit.

Students had a difficult time comprehending the notion of modelling true population by matching fixed data and then deriving the modelling polynomial using arbitrary constants for the data. Often their difficulties stem from not carefully reading what is written. This problem was originally assigned as a one-week project. It probably would be more appropriate to give students two weeks to work on this one.

On the other hand, responses to a survey showed that the students liked this project. The idea of projecting the population (especially the year before the government conducted the census) seemed to capture their attention. Several students commented that they enjoyed seeing a useful application of mathematics.

Finding the Zero of a Polynomial

Consider the polynomial $P(x) = x^3 + 3x - 1$. Your goal is to find a zero of this function: i.e., a number a so that $P(a) = 0$. Although there is an algebraic technique for finding a zero of a cubic polynomial, we are going to approximate a zero. We want the approximation to be within 10^{-2} of an actual zero.

To begin, show that the equation $P(x) = 0$ has at least one solution in the interval $[-1, 1]$. You must give a good justification that such a solution exists.

One way to approximate a solution is to bisect the interval $[-1, 1]$, determine whether $P(x) = 0$ has a solution in $[-1, 0]$ or $[0, 1]$ (as you did in the previous step), and then repeat the process with the new interval containing a solution. How many times must you repeat the bisection process to have a sufficiently accurate answer? Find such an answer. Be sure you understand what it means to bisect an interval.

Calculus gives us another way to perform the search. We use the idea that the tangent line is a good approximation to the graph of a function.

Definition. The line tangent to the curve with equation $y = f(x)$ at a point $(a, f(a))$ on the curve is the line through $(a, f(a))$ with slope $f'(a)$. We call this line a *tangent line*.

Let $y = f(x)$ be a function of x. What is the equation of the line tangent to the graph of f at $(a, f(a))$? What is the x-intercept of this tangent line? Draw a picture to demonstrate what is going on.

Now return to the original problem. Begin with one of the end points of the original interval; this is your original guess. Apply the operation in the previous step, obtaining the x-intercept of the tangent line as a new guess, which hopefully is a better approximation to a solution of the equation $x^3 + 3x - 1 = 0$ than the end point you started with. Is this answer within the desired margin of error from the answer you obtained using the method of bisection? When you get an answer within the margin of error you may stop. Otherwise, repeat the operation, this time beginning with your latest guess.

Compare the two techniques for finding a solution. Which is easier to understand? Why? Which is faster—that is, which leads to an answer within the desired degree of accuracy in the fewest number of iterations?

Extra credit: Instead of using a tangent line, we could use a "tangent quadratic." How would you define the tangent quadratic? Would it work in place of the tangent line? What problems would arise? How do you think it would compare to the two techniques used above?

Comments

Requires: derivatives

Teaches: tangent line, approximation

This problem was assigned when the derivative was first introduced; in fact, the students had this project for a week before we covered the derivative in class. Although this is Newton's method in disguise, no mention is made of Newton's method, so the students got to discover it on their own.

There were two main misconceptions students had with this problem. Several randomly cut the interval up as opposed to actually bisecting it; the third paragraph has been rewritten to help avoid this error. Some applied the margin of error to the values of the polynomial instead of the difference of x values.

Why Astronomers Use Telescopes with Parabolic Mirrors

The star nearest our sun is Alpha Centauri, which is about 4 light years from earth. Alpha Centauri is so far away that when its light reaches earth, it is traveling in essentially parallel rays. To observe distant stars, astronomers use mirrors shaped like paraboloids, which are parabolas rotated about their axes. The reason they use a paraboloidal mirror is that it focuses all the light to a single point, the "focus." (This point is the image of the star in the paraboloidal mirror.) In this project you will demonstrate the focusing property of parabolas.

  a. Suppose our mirror is shaped like the parabola $y = kx^2$, where k is any positive constant. Find the coordinates of its focus and the equation of its directrix in terms of k.

Definition.   The line tangent to the curve with equation $y = f(x)$ at a point $(a, f(a))$ on the curve is the line through $(a, f(a))$ with slope $f'(a)$. We call this line a *tangent line.*

  b. Find the equation of the line tangent to the parabola at a point (x_0, y_0) on the parabola. Then, find the y-intercept of the tangent line.

  c. Consider the triangle formed by a point (x_0, y_0) on the parabola, the y-intercept of the tangent line at this point, and the focus. Draw a picture of this triangle. Prove the triangle is isosceles.

  d. Suppose an incoming light ray strikes a curve at a point (x_0, y_0). If the light ray makes an angle α with respect to the tangent line, then it is reflected at an equal angle to the tangent line. This result from physics is known by the phrase "the angle of incidence equals the angle of reflection." Using this fact, argue that incoming light rays parallel to the axis of the parabola are all reflected to the focus, independent of the point of incidence. Thus, a parabolic mirror focuses incoming light rays parallel to the axis to a point.

  e. The path followed by a ray of light from the star to the focus of the mirror has another special property. Draw a chord of the parabola that is above the focus and parallel to the directrix. Consider a ray of light parallel to the axis as it crosses the chord, hits the parabola and is reflected to the focus. Let d_1 be the distance from the chord to the point of incidence (x_0, y_0) on the parabola and let d_2 be the distance from (x_0, y_0) to the focus. Show that the sum of the distances $d_1 + d_2$ is constant, independent of the particular point of incidence.

Extra credit: Extend your argument from the parabolic cross section to the entire paraboloidal mirror, obtained by rotating this cross section about the y-axis. Thus, prove that all incident rays parallel to the axis of such a mirror focus to a point. Also prove that light traveling along different rays to the focus from a chord perpendicular to the axis will have traveled the same distance, even though these rays were reflected from different points of incidence on the mirror. Thus all of the light waves will arrive in phase, and so interfere constructively to produce a nice bright spot as the image of the star.

Comments

Requires: derivatives, geometry, conic sections

Teaches: tangent line, physics

In this problem students learn about the tangent line, review some geometry, and argue a point logically, while they are doing a problem many find interesting. Students like this project because it is applied. It was assigned as soon as students had seen the derivative, as an introduction to the tangent line.

The Bicycle Race

Carol is a local bicycle racing star and today she is in the race of her life. Moving at a constant velocity k meters per second, she passes a refreshment station. At that instant ($t = 0$ seconds), her support car starts from the refreshment station to accelerate after her, beginning from a dead stop. Suppose the distance travelled by Carol in t seconds is given by the expression kt and distance travelled by the support car is given by the function $\frac{1}{3}(10t^2 - t^3)$, where distance is measured in meters. This latter function is carefully calculated by her crew so that at the instant the car catches up to the racer, they will match speeds. A crew member will hand Carol a cold drink and the car will immediately fall behind.

a. How fast is Carol travelling?

b. How long does it take the support car to catch her?

c. Suppose that Carol is riding at a constant velocity k, which may be different than the value found in part (a). Find an expression for the times when the car and the bike meet which gives these times as a function of her velocity k. How many times would the car and the bike meet if Carol were going faster than the velocity found in part (a)? Slower?

d. Consider a pair of axes with time measured horizontally and distance vertically. Draw graphs that depict the distance travelled by Carol and by the car plotted on the same axes for the original problem (parts (a) and (b)) and for the questions of part (c). You should have three graphs: one for the bike's velocity found in part (a), one for a faster bike, and one for a slower bike. If Carol had been going any faster or slower than the velocity you found in part (a), passing the drink would not have been so easy. Why? Justify your answer.

e. Prove that if a cubic polynomial $P(x)$ has a double root at $x = a$, then $P'(a) = 0$. How does this relate to your answer for part (a) and to your graphs in part (d)?

Extra credit: Generalize part (e) by proving that if $P(x)$ is a polynomial of degree n with a double root at $x = a$, then $P'(a) = 0$.

Comments

Requires: derivatives, product rule

Teaches: velocity

This problem was assigned along with *Finding the zero of a polynomial* (#20) and *Why astronomers use telescopes with parabolic mirrors* (#21). It is not as hard. The students' ability to answer the last question in part (d) seemed to depend on their backgrounds in physics. You might suggest that the student write $P(x)$ in factored form for part (e) and the extra credit.

Equal Areas

Consider the hyperbola $xy = 1$, with $x > 0$ and $y > 0$. We are going to show that the lines tangent to this curve have a very special property.

Definition.   The line tangent to the curve with equation $y = f(x)$ at a point $(a, f(a))$ on the curve is the line through $(a, f(a))$ with slope $f'(a)$. We call this line a *tangent line*.

The tangent line at any point on the hyperbola forms a right triangle with the coordinate axes. Your first problem is to show that the area of this triangle does not depend on the point chosen. To prove this, you may want to proceed as follows. Expressing y as a function of x, you are led to consider the function $H(x) = \frac{1}{x}$. What is the equation of the line tangent to the graph of H at $(a, H(a))$? Draw a graph of this function, a typical tangent line, and the triangle in question. What is the area of the triangle? If the area depends on the point $(a, H(a))$ you have made a mistake.

Now comes the interesting part. Note that the function $H(x) = \frac{1}{x}$ has the property that $H'(x) = \frac{-H(x)}{x}$. Let f be any function for which $f(x) > 0$ whenever $x > 0$ and the derivative $f'(x)$ is always negative. Suppose that the derivative of f satisfies the equation $f'(x) = C\frac{f(x)}{x}$, where C is a negative constant. Give several examples of functions f that satisfy these properties. Consider a triangle formed by the coordinate axes and the line tangent to the graph of f at a point $(a, f(a))$. Assume that it has area that is independent of the point $(a, f(a))$. Prove that f is a constant multiple of H. Which of your examples satisfy all of these properties? Which of your examples do not satisfy the constant area property?

Hint: To show this last part, you can use the condition on the area of the triangle to get a relation involving f and f'. Then recall the condition we are assuming on the derivative of f.

Comments

Requires: derivatives

Teaches: tangent line, arbitrary constants

This project has never been assigned.
This is a very nice problem, but students may find the solution to the first part in calculus textbooks or solutions manuals. The second part involves algebra with arbitrary constants; the students are asked to work with tangent lines at arbitrary points on the graph of f. This is often quite challenging for the students.

Invariant Areas

Consider the hyperbola $xy = 1$, with $x > 0$ and $y > 0$. We are going to show that the lines tangent to this curve have a very special property.

Definition.   The line tangent to the curve with equation $y = f(x)$ at a point $(a, f(a))$ on the curve is the line through $(a, f(a))$ with slope $f'(a)$. We call this line a *tangent line.*

The tangent line at any point on the hyperbola forms a right triangle with the coordinate axes. Your first problem is to show that the area of this triangle does not depend on the point chosen. To prove this, you may want to proceed as follows. Expressing y as a function of x, you are led to consider the function $H(x) = \frac{1}{x}$. What is the equation of the line tangent to the graph of H at $(a, H(a))$? Draw a graph of this function, a typical tangent line, and the triangle in question. What is the area of the triangle? If the area depends on the point $(a, H(a))$ you have made a mistake.

Now comes the interesting part. You are going to find all functions that have this property. Let f be any function such that $f(x) > 0$ whenever $x > 0$, the derivative $f'(x)$ is always negative, and the second derivative $f''(x)$ is never zero. Consider a triangle formed by the coordinate axes and the line tangent to the graph of f at a point $(a, f(a))$. Assume that it has area that is independent of the point $(a, f(a))$.

As in the first part, find an explicit formula for the line tangent to the curve $y = f(x)$ at the point $(a, f(a))$. Then, find the area of the triangle formed by the tangent line and the coordinate axes.

Assume that the area is constant, which gives you an equation in the variable a. Differentiate this equation in a, simplify, factor, and use the conditions on f to arrive at a simple differential equation. Solve this differential equation to find all the solutions to the problem.

Comments

Requires: derivatives

Teaches: tangent line, differential equations

This project has never been assigned.

Although students may find the solution to the first part of this problem in calculus textbooks or solutions manuals, the second part is interesting and not in textbooks. The students are asked to work with tangent lines at arbitrary points on the graph of f. This is often quite challenging for the students.

If one follows the directions in the last paragraph, one gets the equation

$$f''(a)[f(a) - af'(a)][f(a) + af'(a)] = 0.$$

The conditions on f reduce this to $f(a) + af'(a) = 0$. Since the left-hand side is a derivative, students can solve it without needing the natural logarithm.

A New Derivative

New advances in mathematics occur every day and one of the most recent (and least useful) is a new derivative $D * f$.

Definition.   A function f is called *-*differentiable* at a if
$$\lim_{x \to a} \frac{(f(x))^2 - (f(a))^2}{x - a}$$
exists. If the limit exists, we call it the *-*derivative* of f at a and denote it by $D * f(a)$.

You are going to develop some of the properties of the *-derivative. In parts (a), (b), and (c) below you may use properties of limits, but do not use any properties of the derivative you learned about in class. However, you may wish to read the text before tackling your tasks.

a. Prove the following theorem:

Theorem.   *If f is *-differentiable at a, then f^2 is continuous at a.*

Note that you are not being asked to prove that f is continuous at a. In fact, see part (d) below.

b. State and prove a product rule for the *-derivative. That is, complete the statement of the following theorem and then prove it.

Theorem (Product rule).   *If f and g are *-differentiable at a, then $f \cdot g$ (the product of f and g) is *-differentiable at a and*
$$D * (f \cdot g)(a) = \text{(complete the formula)}.$$

c. State and prove a quotient rule for the *-derivative. You will need to decide what hypotheses are needed and what the formula for the *-derivative of the quotient will be.

d. Find a function that is *-differentiable at some point a but not continuous at a. Prove that your example has the indicated properties.

e. Find a function that is *-differentiable at some point a, continuous at a, but not differentiable at a. Prove that your example has the indicated properties.

f. Find two functions f and g that are * differentiable at some point a but $f + g$ is not *-differentiable at a. Justify.

g. Finally, state and prove a theorem that relates $D * f(a)$ and $f'(a)$. You will need
    to decide what hypotheses are necessary.

Comments

Requires: limits, derivatives, continuity, product rule, quotient rule

Teaches: derivatives, continuity, product rule, quotient rule

   This problem was assigned in an accelerated calculus class for above average students. They did fairly well, but thought it was hard. They needed help in finding the
examples required in parts (d), (e), and (f). Most of them used the statements and
proofs of the product rule and quotient rule in the text to solve parts (b) and (c).
   Similar, but different, projects could be created by using a different power or
other function in the definition of the *-derivative.

Dos Moscas

Two flies (twin sisters) are sitting on a spherical balloon while it is being inflated at a constant rate. Assume that air is being injected into the balloon at a rate of 5 cubic centimeters per second, and that the balloon has no air in it to begin with. Further assume that one sister is situated at the north pole and the other on the equator.

a. Draw a picture complete with labels.

b. How fast are the two flies parting company as a function of time?

c. How fast after 3 seconds?

d. How fast after 1/10 second?

e. How fast after 1/100 second?

f. How fast are they separating initially?

Comments

Requires: derivatives

Teaches: related rates

This project was designed for and used in a calculus course for biology and social science majors. They had not yet studied related rates. The project should be assigned before implicit differentiation.

Note that there are two metrics, the distance along a chord or on the sphere, that can reasonably be used to measure the distance between the flies.

The Sun as a Black Hole

The acceleration of a particle due to the force of gravity exerted upon it by a ball is directly proportional to the mass of the ball, and inversely proportional to the square of the distance from the particle to the center of the ball. Find and study a derivation of escape velocity from a ball in a gravitational field. The escape velocity is the minimum velocity needed to escape from the surface of the ball.

Scientists call a black hole an object in space from which no light can escape. Suppose chemist Marie Curie wonders if our sun could be compressed to become a black hole, and if so, how small it would need to be. Although she has no time to spare to find the answer, she hires you to solve it! What would you report? For your analysis you may assume that the result of the escape velocity calculation for bodies with mass also holds for massless photons of light, and that an object becomes a black hole when the escape velocity exceeds the speed of light.

You may wish to use the following data. At the surface of the earth, which has radius 6,436 kilometers, the acceleration due to gravity is -9.8 meters/sec^2. The sun is 332,000 times more massive than the earth, and the speed of light is 299,274 kilometers/sec.

Extra credit: Discuss the same problem assuming that the force of gravity varies inversely as:

a. the $\frac{3}{2}$ power of the distance.

b. the $\frac{1}{2}$ power of the distance.

Comments

Requires: derivatives, arbitrary constants

Teaches: arbitrary constants, physics

Our students based their work on a derivation of escape velocity given in their text (*Calculus and Analytic Geometry,* Sherman K. Stein, 4th edition, McGraw-Hill). Students may need some similar reference to complete the project. They also may need to read about the meanings of direct and inverse proportionality.

A Road Between Two Towns

Two towns lie to the south of a straight road, but unfortunately they are neither connected to it, nor to one another. The citizens of the two towns decide to build two roads, one from each town, to the existing road. They are cost conscious and are smart enough to hire a calculus student like yourself as a consultant to choose the route for the roads.

a. Use calculus to find the shortest route connecting the two towns via the existing road.

b. Use elementary geometry to solve the same problem.

   Hint: Suppose the two towns were on opposite sides of the existing road.

   Now you are going to use your solution to the road problem in a very surprising way to solve an old familiar problem.

c. Let A and B be foci of an ellipse, P a point on the ellipse, and L the line tangent to the ellipse at P. If you think of A and B as the towns and L as the existing straight road described above, show that P is the point where the shortest route from A to L and back to B connects with L. Combine this with your argument in (b) to conclude that the line segments $\overline{AP}$ and $\overline{BP}$ form equal angles with L.

d. What you have proven is the reflective property of the ellipse. Do some reading and write a paragraph describing a practical use of this property.

Extra credit: Are there ever situations where there is a cheaper way to connect the two towns and the road? Analyze the situation.

Comments

Requires: one-variable extrema, derivatives of trigonometric functions, geometry, conic sections, tangent line

Teaches: conic sections, one-variable extrema

   As originally assigned, the project consisted of parts (a) and (b). The solution may be in some calculus books. The part concerning the ellipse was added and it also may be found in some books. References to the reflective property of the ellipse occur in many precalculus and calculus texts.

Dome Tent

Imagine making a tent in the shape of a spherical cap (a sphere with a lower portion sliced away by a plane). Assume we want the volume to be $2.2m^3$, to sleep two or three people. Draw a picture, identifying all appropriate variables. The floor of the tent is cheaper material than the rest: assume that the material making up the dome of the tent is 1.4 times as expensive per square meter than the material touching the ground.

   a. What should the dimensions of the tent be so that the cost of the material used is a minimum?
   b. What is the total area of the material used?

   Now change the problem so that the floor of the tent is more expensive material than the rest: assume that the material touching the ground is 1.4 times as expensive per square meter than the material making up the dome of the tent.

   c. What should the dimensions of the tent be so that the cost of the material used is a minimum?
   d. What is the total area of the material used?
   e. How practical would these two tents be?

Pup Tent

Imagine making a tent in the shape of a right prism whose cross section is an equilateral triangle (the door is on one of the triangular ends). Assume we want the volume to be 2.2m^3, to sleep two or three people. Draw a picture, identifying all appropriate variables. The floor of the tent is cheaper material than the rest: assume that the material making up the ends and top of the tent is 1.4 times as expensive per square meter than the material touching the ground.

   a. What should the dimensions of the tent be so that the cost of the material used is
      a minimum?
   b. What is the total area of the material used?

   Now change the problem so that the floor of the tent is more expensive material than the rest: assume that the material touching the ground is 1.4 times as expensive per square meter than the material making up the ends and top of the tent.

   c. What should the dimensions of the tent be so that the cost of the material used is
      a minimum?
   d. What is the total area of the material used?
   e. How practical would these two tents be?

Comments on *Dome Tent* and *Pup Tent*

Requires: one-variable extrema

Teaches: one-variable extrema

   The shapes are unexpected and a bit strange; you may want to change the parameters. The students need to look up formulae in a handbook for the volume of the spherical cap. This has only been used in the calculus course for students in biology and the social sciences.
   These two projects, *Dome tent* (#29) and *Pup tent* (#30), may appear to be the first members of a parameterized series of projects about tents of various shapes: rectangular prisms, pyramids, hexagonal cylinders, etc. But be careful! They vary enormously in difficulty. For example, the pyramid on a square base, with the same questions asked, is much harder than the tent problems given here.

A Greenhouse Extension

Your parents are going to knock out the bottom of the entire length of the south wall of their house, and turn it into a greenhouse by replacing some bottom portion of the wall by a huge sloped piece of glass (which is expensive). They have already decided they are going to spend a certain fixed amount. The triangular ends of the greenhouse will be made of various materials they already have lying around.

The floor space in the greenhouse is only considered usable if they can both stand up in it, so part of it will be unusable, but they don't know how much. Of course this depends on how they configure the greenhouse. They want to choose the dimensions of the greenhouse to get the most usable floor space in it, but they are at a real loss to know what the dimensions should be and how much usable space they will get. Fortunately they know you are taking calculus. Amaze them!

Comments

Requires: derivatives of trigonometric functions, one-variable extrema

Teaches: arbitrary constants

This project was surprising to students since at first sight it contains no overt mathematics. It teaches students to determine arbitrary constants, choose an independent variable, and determine other quantities as a function of the independent variable.

Calculus in the Courtroom

The great metropolis of Lesser Anytown, USA, is famous for its long curved boulevard in the shape of a perfect parabola which passes through a large rectangular park.

Late one night, Officer James "Handcuff" Put'emaway is admiring the beautiful fountain that marks the focus of the parabola, located directly northward on his left just as he cruises along the boulevard across the vertex of the parabola. At that moment he notices through the trees a suspicious vehicle moving at constant speed in an easterly direction on the street that bounds the park along its southern edge. He is immediately able to determine the vehicle's speed and how far east of him it is, by using his watch and the known distance between street lights along the street. He continually adjusts his own speed along the boulevard so that he remains always the same distance west of the suspect car, and trains his infrared spotlight on the driver. He is able to observe the driver's actions for a while, by turning his spotlight to keep it on the driver. But several seconds before his spotlight would be pointing directly sideways, he finds that it is stuck and will not pivot any farther. Nevertheless, at this time he feels he has enough information for an arrest.

Later, when Handcuff is subpoenaed to testify at the trial, he returns to the scene to measure the shortest distance from the street to his location when he first observed the suspect, thinking he might need this information in court. After he has given all the information he has on the witness stand, the defense attorney, Julie "Lightning" Get'emoff, asks, "And just how long did you have your spotlight on the driver while you observed all this? Did you time it?" Handcuff answers, "No, I didn't clock it, but it was quite a while!" She says to the judge, "Your Honor, if we can recess for lunch now so that I can examine the officer's patrol car, I think I can show the court how to answer this very important question. As we all know, the law requires at least 20 seconds of observation under these circumstances, before an arrest is allowed. And I believe the evidence will show that there was insufficient observation time for the officer to justify an arrest!"

After lunch, Julie returns with a calculus student named Flora as an expert witness. On the witness stand, Flora says, "Your honor, suppose that, in addition to the evidence that has already been presented to the court, we were to know the distance from that fountain to where Officer Put'emaway began his observations. The unknown quantity I wish to determine is the length of time the officer observed the suspect using his spotlight. Now you can see from my work on these pages that using a little calculus, we can obtain an equation involving a third degree polynomial in the unknown time. So we can just solve for the unknown." The judge replies, "But doesn't a third degree polynomial have as many as three roots? How will we know which one we need?" Flora responds, "Well, your Honor, here on this page I have proved using some calculus that this equation has to have a unique positive solution." The judge says, "Good, and what is the solution, Flora?" "Well, your Honor, I can approximate the solution using a technique from calculus known as Newton's method. As my first approximation I'll just pick zero, and then the method gives this number as my second approximation. But at some point soon, your Honor, we'll really need to know the distance to that fountain to learn anything useful." The judge recesses the trial again to have the L.A.P.D. (Lesser Anytown Parks Department) find the relevant distance.

During the recess, Julie is concerned, and says to Flora, "Gosh, I expected we'd get an exact solution for the length of observation time. If you only have approximations to the true solution, how will I ever be able to demonstrate for sure that Handcuff observed for too short a time?" "Oh, don't worry about that," Flora replies, "because I've proved using calculus again on this page here that for our equation, the true solution has to be smaller than this second approximation. So if the Parks Department reports back that the distance to the fountain is less than the number I've written here, then we've already won the case based on my work so far. And if not, then we still might win it if I carry Newton's method a few more steps." "Wow," says Julie, "I guess you're right."

Pretend that you are Flora the calculus student. Present the mathematics and explanations that back up all five of her claims.

Extra credit: Later on at home, Flora's roommate Vivian says, "But Flora, if all you want to know is whether Handcuff observed long enough, and you don't care to really know how long he observed for, then there's a much easier way than all that work with Newton's method, since we know quite a bit about the behavior of the function in your equation from calculus." Flora replies, "Yes, I guess you're right, but the judge was such a stickler. He really did want to know more about how long Handcuff eyed that guy. That's why I had to use Newton's method to approximate the actual solution for the judge."

Explain what Vivian's method was.

Comments

Requires: mean value theorem, intermediate value theorem, geometry, trigonometry, arbitrary constants

Teaches: Newton's method, arbitrary constants

Students liked this project and were quite successful with it. They worked particularly well on it in groups. It involves a good bit of geometry and trigonometry, and it takes about half the time just to get started.

The Shortest Crease

You have a sheet of paper that is 6 units wide and 25 units long, placed so that the short side is facing you. Fold the lower right corner over to touch the left side. Your task is to fold the paper in such a way that the length of the crease is minimized. What is the length of the crease?

Comments

Requires: geometry, one-variable extrema, trigonometry

Teaches: one-variable extrema

This is a good example of how writing can affect a project. The problem is simple to state and, with a little experimentation, to understand. The calculus is fairly simple (using the chain rule and the quotient rule). The hard part of the problem is finding a function to describe the length of the crease. This involves a bit of geometry or trigonometry and some imagination. It is difficult to see how to give hints without telling the students how to solve the problem.

This is probably a one-week project.

The Instantaneous Value of the Bending Ratio

After your first semester of calculus, you are lucky enough to land a summer job working for NASA because of your knowledge of the mean value theorem. Given a proposed flight path for a manned rocket, your job is to determine about how rough or smooth the flight may seem to someone on board. Your employer has certain criteria that measure these quantities. The first is the angle between the x-axis and the line tangent to the flight path, which should not change too fast. If the rocket travels in a straight line, say $y = mx + b$, where m and b are constants, then the angle between the line tangent to the graph of $y = mx + b$ and the x-axis is the same, no matter where the rocket may be.

a. Compute the angle between the x-axis and the line tangent to the curve $y = mx+b$ at any point. Do you get the same answer for different points of tangency?

Certainly travelling in a straight line would be considered a smooth ride. Use your answer to part (a) to justify this statement. If this angle does change, then the flight path is curved, and the ride is a little rougher. Your supervisor, Dr. Rockette, realizes this and has already computed some function $A(t)$, where t represents time, that gives the angle between the tangent to the flight path and some fixed axis. She then explains that part of your job is to determine how quickly $A(t)$ is changing. If the angle $A(t)$ is changing quickly, then the direction the rocket is headed is changing quickly as well.

At this point you interrupt and question whether knowing the change in $A(t)$ is enough information. After all, you expound, if the angle has changed 90°, then this could be a gradual 90° turn or a sharp 90° turn. We need to distinguish between these two cases. You continue that for a gradual 90° turn, a lot of distance is covered when making the turn, while very little distance is covered for a sharp turn. If we knew the distance travelled, then we could compare the change in the angle with the change in the distance. Impressed with these remarks, Dr. Rockette explains that her engineering team has considered the distance function. However, they refer to this as arc length, $L(t)$, since the distance travelled is the length of the arc of the rocket path. They have decided to consider the "bending ratio," which is the change in the angle $A(t)$ with respect to arc length. In other words, we would like to know the value of

$$B(t_1, t_0) = \frac{A(t_1) - A(t_0)}{L(t_1) - L(t_0)},$$

where t_0 and t_1 are two time values that are close to each other. For your calculations of the bending ratio, you will use the following lemma.

Lemma. *If $A(t)$ and $L(t)$ are continuous on $[t_0, t_1]$, differentiable on (t_0, t_1), and $L(t_1) \neq L(t_0)$, then there is some constant c with $t_0 < c < t_1$, and*

$$\frac{A'(c)}{L'(c)} = \frac{A(t_1) - A(t_0)}{L(t_1) - L(t_0)}.$$

b. Prove this lemma.

Hint: Review the proof of the mean value theorem and try to construct a similar proof for this lemma.

We would like to find the instantaneous value of the bending ratio, i.e., $\lim_{t_1 \to t_0} B(t_1, t_0)$. Suppose A and L have continuous derivatives and L' is never 0.

c. Using the above lemma, find a general formula for

$$\lim_{t_1 \to t_0} B(t_1, t_0).$$

d. Apply this general formula to the specific functions

$$A(t) = \alpha_1 \sin(t) + \alpha_2 \quad \text{and} \quad L(t) = \alpha_3 t^2 + \alpha_4 t + 1,$$

and compute $\lim_{t \to 0} B(t, 0)$ in this case. You may assume that α_1, α_2, α_3, and α_4 are positive constants and $t \geq 0$.

Extra credit: Suppose that $A(t) = \frac{\pi}{2} \cos(t)$ and $L(t) = t^2$. Find $\lim_{t \to 0} B(t, 0)$. Is this realistic for a rocket fired from a launch pad at $t - 0$?

Estimating Bounds for the Bending Ratio

After your first semester of calculus, you are lucky enough to land a summer job working for NASA because of your knowledge of the mean value theorem. Given a proposed flight path for a manned rocket, your job is to determine about how rough or smooth the flight may seem to someone on board. Your employer has certain criteria that measure these quantities. The first is the angle between the x-axis and the line tangent to the flight path, which should not change too fast. If the rocket travels in a straight line, say $y = mx + b$, where m and b are constants, then the angle between the line tangent to the graph of $y = mx + b$ and the x-axis is the same, no matter where the rocket may be.

a. Compute the angle between the x-axis and the line tangent to the curve $y = mx + b$ at any point. Do you get the same answer for different points of tangency?

Certainly travelling in a straight line would be considered a smooth ride. Use your answer to part (a) to justify this statement. If this angle does change, then the flight path is curved, and the ride is a little rougher. Your supervisor, Dr. Rockette, realizes this and has already computed some function $A(t)$, where t represents time, that gives the angle between the tangent to the flight path and some fixed axis. She then explains that part of your job is to determine how quickly $A(t)$ is changing. If the angle $A(t)$ is changing quickly, then the direction the rocket is headed is changing quickly as well.

At this point you interrupt and question whether knowing the change in $A(t)$ is enough information. After all, you expound, if the angle has changed $90°$, then this could be a gradual $90°$ turn or a sharp $90°$ turn. We need to distinguish between these two cases. You continue that for a gradual $90°$ turn, a lot of distance is covered when making the turn, while very little distance is covered for a sharp turn. If we knew the distance travelled, then we could compare the change in the angle with the change in the distance. Impressed with these remarks, Dr. Rockette explains that her engineering team has considered the distance function. However, they refer to this as arc length, $L(t)$, since the distance travelled is the length of the arc of the rocket path. They have decided to consider the "bending ratio," which is the change in the angle $A(t)$ with respect to arc length. In other words, we would like to know the value of

$$B(t_1, t_0) = \frac{A(t_1) - A(t_0)}{L(t_1) - L(t_0)},$$

where t_0 and t_1 are two time values that are close to each other. For your calculations of the bending ratio, you will use the following lemma.

Lemma. *If $A(t)$ and $L(t)$ are continuous on $[t_0, t_1]$, differentiable on (t_0, t_1), and $L(t_1) \neq L(t_0)$, then there is some constant c with $t_0 < c < t_1$, and*

$$\frac{A'(c)}{L'(c)} = \frac{A(t_1) - A(t_0)}{L(t_1) - L(t_0)}.$$

b. Prove this lemma.

   Hint: Review the proof of the mean value theorem and try to construct a similar proof for this lemma.

c. You are to estimate the bending ratio during the first second of the flight. Suppose that

$$A(t) = \alpha_1 \sin(t) + \alpha_2 \quad \text{and} \quad L(t) = \alpha_3 t^2 + \alpha_4 t + 1,$$

   where α_1, α_2, α_3, and α_4 are positive constants. Use the above lemma to find upper and lower bounds for $B(t, 0)$ where $0 < t \leq 1$.

Extra credit: Suppose that $A(t) = \frac{\pi}{2} \cos(t)$ and $L(t) = t^2$. Find reasonable upper and lower bounds on $B(t, 0)$ where $0 < t \leq 1$. Are these bounds ever realized?

Comments on *The Instantaneous Value of the Bending Ratio* and *Estimating Bounds for the Bending Ratio*

Requires: Rolle's theorem

Teaches: mean value theorem, estimation

These projects require the students to use Rolle's theorem. It is important to get them to be precise in the use of a theorem. Many did not distinguish between the hypothesis and the conclusion.

The pedagogical goal of these projects is to get the students to prove the lemma in part (b). The lemma is not necessary for part (c) of *The instantaneous value of the bending ratio* (#34). Many students had difficulty using the maximum of the numerator and the minimum of the denominator to estimate the size of a fraction in *Estimating bounds for the bending ratio* (#35).

Finding Your Area

In this project, you are going to accurately trace the shape of a member of your group and approximate the area of the picture that you create. Your main tasks are to devise a method for approximating the area and to show that your approximation is very close to the actual area.

To begin, start with a sheet of paper that is seven feet by three feet. Have one member of the group lie on the paper and make an accurate trace of the outline of that person. You will have to hand in this paper with your project.

Devise a method to approximate the area of the region inside the curve you have traced. You must explain your method in detail and why it works.

Using the method devised above, find an approximation of the area that differs from the actual area by less than 1% of the actual area. This is an important part of the project and you should spend some time thinking about how accurate an answer is required. Reading about "relative errors" might be of help. You will not know the actual area, so you must devise a way of showing that your approximation is within the desired degree of accuracy.

Finding the Area of Your Hand

In this project, you are going to accurately trace the shape of your hand and approximate the area of the picture that you create. Your main tasks are to devise a method for approximating the area and to show that your approximation is very close to the actual area.

To begin, start with a sheet of blank paper and make an accurate trace of the outline of your hand. You will have to turn in this paper with your project.

Devise a method to approximate the area of the region inside the curve you have traced. You must explain your method in detail and why it works.

Using the method devised above, find an approximation of the area that differs from the actual area by less than 1% of the actual area. This is an important part of the project and you should spend some time thinking about how accurate an answer is required. Reading about "relative errors" might be of help. You will not know the actual area, so you must devise a way of showing that your approximation is within the desired degree of accuracy.

Comments on *Finding Your Area* and *Finding the Area of Your Hand*

Requires: introduction to integration

Teaches: approximating areas, error estimation

The first project, *Finding your area* (#36), was assigned to students working in groups of two or three. It was assigned when the students were first seeing the definite integral as a limit of Riemann sums. The idea of the project is to get students to work on geometric ideas related to integration.

The last part of the problem deals with approximation and relative error. Students had a lot of difficulty with this because they have trouble with absolute values and inequalities.

Two groups chose to cut out the outline of the body and weigh it with very accurate scales. To establish constant density per unit area, one group used a hole punch to punch 30 holes in the paper and weighed each of the 30 discs of paper. They fondly dubbed the outline of the body that they turned in "Shotgun Willie." If you do not want your students to approach the problem this way, you should add a note to that effect.

The project involves a lot of busy work, so if you want an individual project or a one-week project, you might try the second project instead. It requires less time since the region to be measured is much smaller.

$\int_1^x \frac{1}{t} dt$ Should Equal $\log x$

When we consider the integrals of t^r, all the antiderivatives are of the same form except for $r = -1$. This is often confusing, even though the result for $r = -1$ is consistent with all the other results. In this problem, you are going to prove this is the case. In order to complete this project, you will need to be familiar with logarithms and l'Hôpital's rule.

a. Let $F(r) = \int_1^2 t^r dt$, with r a real number not equal to -1. Evaluate $F(r)$ for several values of r close to -1. You will probably want to use a calculator for this part. Do the values of $F(r)$ seem to approach a limit? (You must choose enough values of r to see a pattern to be able to answer this question.) Do you recognize the limit? Replace the upper limit 2 by 3, 4, and 10 and answer the same questions in each case.

b. Let b be a fixed positive number. For r a real number not equal to -1, redefine the function F by $F(r) = \int_1^b t^r dt$. Find a simpler formula for $F(r)$.

c. Show that F is a continuous function. Give a good justification for your answer.

d. How should we define $F(-1)$ so that F is continuous at -1? Show that this value makes F continuous at -1. Reread the introductory paragraph and explain the title of this project.

e. Explain your results from part (a) in light of your discoveries in part (d).

Comments

Requires: integration, natural logarithm, l'Hôpital's rule

Teaches: continuity, natural logarithm

This problem attempts to give the students a way to see that the integral of $\frac{1}{t}$ is consistent with the integration of other exponents. The problem is a little circuitous, although the students will probably not see that.

Unfortunately, the students seemed to miss the point of this project. They did not really understand how to answer the last part of part (d), and they did not take the introduction into account. Part (a) has been added on account of this, as well as the comment in part (d) about rereading the introduction.

Upper and Lower Riemann Sums

Let $f(x)$ be a positive, continuous, increasing function on $[a, b]$. Your goal is to develop a technique to estimate the integral $\int_a^b f(x)dx$ within a given margin of error without evaluating the integral.

a. Your first step is to understand two special kinds of Riemann sums. Partition $[a, b]$ into n equal parts. What is the length of each subinterval? For the first kind of Riemann sum, choose the height of each rectangle to be the maximum value of f over the subinterval. How do you know the maximum exists? For what value of x do you get the maximum value? The Riemann sum defined by taking the maximum value of f over each subinterval is called an *upper sum*. Explain why $\int_a^b f(x)dx$ is less than or equal to any upper sum. Write explicitly the upper sum when $n = 4$.

b. For the second kind of Riemann sum, repeat part (a), but replace the maximum value of f over each subinterval by the minimum value. This Riemann sum is called a *lower sum*. Explain why $\int_a^b f(x)dx$ is greater than or equal to any lower sum. Write explicitly the lower sum when $n = 4$.

c. Let U_4 be the upper sum and L_4 the lower sum for $n = 4$ found above. The difference $U_4 - L_4$ is nonnegative. Why? Simplify $U_4 - L_4$ as much as possible. Next, find a very simple expression for $U_n - L_n$. Explain why your expression is correct.

d. Part (c) gives you a way to make the upper and lower sums as close to each other as you wish. Suppose ϵ is a given positive real number. How can you use the result of part (c) to estimate $\int_a^b f(x)dx$ with a margin of error at most ϵ? Explain.

e. Estimate $\int_{0.5}^2 (1 + x^2)^{1/3}dx$ with an error at most 10^{-1} using the ideas from part (d).

Extra credit: Can you think of other ways (other than the maximum or minimum) to choose the value of f on each subinterval that would give better approximations to the integral than U_n or L_n? Discuss your ideas.

Comments

Requires: integration

Teaches: Riemann sums, approximating integrals

Students had a difficult time with this project and it has been extensively rewritten. The goals are for them to get a real feel for Riemann sums and how they approximate integrals. This could be used to help them understand the definition of the integral as a limit of Riemann sums.

Many students had a hard time seeing how to use the information that U_n is close to L_n in order to estimate the integral. In general, they forgot that from parts (a) and (b) they had discovered that

$$L_n \leq \int_a^b f(x)dx \leq U_n.$$

Instructors should be aware that students will have to be reminded of this point.

Houdini's Escape

Harry Houdini was a famous escape artist. In this project we relive a trick of his that challenged his mathematical prowess, as well as his skill and bravery. It will challenge these qualities in you as well.

Houdini had his feet shackled to the top of a concrete block which was placed on the bottom of a giant laboratory flask. The cross-sectional radius of the flask, measured in feet, was given as a function of height z from the ground by the formula

$$r(z) = \frac{10}{\sqrt{z}},$$

with the bottom of the flask at $z = 1$ foot. The flask was then filled with water at a steady rate of 22π cubic feet per minute. Houdini's job was to escape the shackles before he was drowned by the rising water in the flask.

Now Houdini knew it would take him exactly ten minutes to escape the shackles. For dramatic impact, he wanted to time his escape so it was completed precisely at the moment the water level reached the top of his head. Houdini was exactly six feet tall. In the design of the apparatus, he was allowed to specify only one thing: the height of the concrete block he stood on.

a. Your first task is to find out how high this block should be. Express the volume of water in the flask as a function of the height of the liquid above ground level. What is the volume when the water level reaches the top of Houdini's head? (Neglect Houdini's volume and the volume of the block.) What is the height of the block?

b. Let $h(t)$ be the height of the water above ground level at time t. In order to check the progress of his escape moment by moment, Houdini derives the equation for the rate of change $\frac{dh}{dt}$ as a function of $h(t)$ itself. Derive this equation. How fast is the water level changing when the flask first starts to fill? How fast is it changing when the water just reaches the top of his head? Express $h(t)$ as a function of time.

c. Houdini would like to be able to perform this trick with any flask. Help him plan his next trick by generalizing the derivation of part (b). Consider a flask with cross sectional radius $r(z)$ (an arbitrary function of z) and a constant inflow rate $\frac{dV(t)}{dt} = A$. Find $\frac{dh}{dt}$ as a function of $h(t)$.

Extra credit: How would you modify your calculations to take into account Houdini's volume, given Houdini's cross-sectional area as a function of height?

Comments

Requires: volume by slices, second fundamental theorem of calculus, chain rule, natural logarithm

Teaches: second fundamental theorem of calculus

The students seemed to enjoy this one. They viewed it as an "applied" problem!

Commuting Powers

Addition is commutative, and so is multiplication. But exponentiation is not. The object of this project is to investigate those unusual pairs of *positive* numbers, x and y, for which

$$x^y = y^x.$$

Try to find some such pairs. Can you find some when $x \neq y$?

To find more pairs in a systematic fashion, fix a value of y and use Newton's method and a computer or pocket calculator to find a corresponding value for x . Plot enough of these pairs (x, y) to see the general pattern of the locus in the first quadrant. There should be two distinct values. Show how the initial starting value for an application of Newton's method affects which curve the resulting point lies on.

Where do the two curves of the locus intersect? To get a start in answering this, separate the variables by taking the xth root and then the yth root of both sides of the equation above. Either side of this equation gives a new function to study. Now graph the function, $u = t^{1/t}$. With the shape of the graph as a guide and a little thought, use the calculus to answer this question rigorously.

There is another way to study this problem. Introduce a new variable:

$$t = \frac{y}{x}.$$

Find expressions for x and y as functions solely of t. Use this parametric representation to begin to fill out your original locus. Which part of the locus do the parametric equations yield? Support this with a mathematical argument. For what value of t do you get the other part? How can you use this representation to find the point of intersection in another way by taking the limit of t at a critical value?

Comments

Requires: parametric equations, exponential functions, Newton's method, computer or programmable calculator

Teaches: noncommutativity, parametric equations, Newton's method

This project has never been assigned.

It would appear difficult, if not impossible, at first glance, to be able to say much about when powers commute. This project shows that there are at least three ways to go about this, including one that goes back to Euler. This project is hard, requiring more sophisticated analysis than typical students can provide.

Integration by Parts Using Constants

With integration by parts given by the formula

$$\int u\,dv = uv - \int v\,du,$$

when we find v, we generally do not include a constant of integration. For example, to integrate $\int xe^x\,dx$ we may set $u = x$ and $dv = e^x\,dx$, and then use $v = e^x$ instead of $v = e^x + C$. This problem explores whether or not it is justified to leave off the constant of integration and also whether it could be a benefit to include a constant of integration.

a. Consider the integral $\int x\cos(x)\,dx$. Evaluate this integral three different ways using integration by parts. The first time do it in the usual way. Next evaluate the integral again, but this time include the constant of integration "+1." The last time, include the constant of integration "+C." Compare the answers and explain any differences. Which was the easiest technique?

b. Consider the integral $\int x\arctan(x)\,dx$. Evaluate this integral using integration by parts in the usual way. Then find a constant of integration that makes the last integration trivial. Compare the answers and explain any differences. Which was the easiest technique for this integral?

c. Make a conjecture relating integration by parts with and without using the constant of integration. Your conjecture should have hypotheses and a conclusion and involve arbitrary functions u and v.

d. Prove that your conjecture is correct.

Comments

Requires: integration by parts, inverse trigonometric functions

Teaches: integration by parts

This problem turned out to be too easy for a two-week project; most students could even figure out and outline a proof with little or no help. This might be appropriate as a one-week assignment.

Exploring Uncharted Waters

You have taken a summer job with Jock Crusto, a world famous oceanographer. His latest endeavor is to explore the uncharted Specific Ocean on the newly discovered planet Mickey. The exploration will involve three ships, the Pinto, the Lima, and the Verde. All three ships will leave Port Minnie at the same time. Set up a coordinate system with the origin at Port Minnie, the positive y-axis pointing north and the positive x-axis pointing east. The lead ship, the Pinto, will follow a parabolic path along the curve $y = x^2$ at a speed of 15 knots. The second ship, the Lima, will travel in a northeasterly direction along a parabolic path such that it is always due south of the Pinto. The third ship will also travel in a northeasterly direction along a parabolic path such that it is always due west of the Pinto. For the Lima and Verde, the parabolas both have their vertices at the origin. In addition, the ships must travel in such a way that the line between the Lima and the Pinto and the line between the Verde and the Pinto sweep out equal areas in equal time. Each ship has a computer-driven guidance system so there is little for you to do but watch the monitors showing pictures of the ocean bottom.

But wait!!! Before leaving Port Minnie, a last minute check of the computers reveals a serious problem. All is well with the Pinto and the Lima, but the guidance software for the Verde is destroyed. There are just a few hours to correct the situation or the mission will have to be postponed for months.

There is a computer programmer on board who can reprogram the Verde's computer, but someone needs to determine the course for the Verde. Jock knows that you know some calculus, so you are asked to give the programmer the x and y components for the motion of the Verde. One bit of information that you need is that, when the Pinto is $3\sqrt{10}$ nautical miles (straight line distance) from Port Minnie, the Lima is 1.8 nautical miles due south of the Pinto.

Comments

Requires: speed, parametric equations, area, second fundamental theorem of calculus

Teaches: second fundamental theorem of calculus, speed

A more theoretical version without the story line and the questions about speed and components of velocity has been given several times before. The added twists make this into a nice two-week project.

Integration by the Method of Undetermined Coefficients

There are many integration problems that are straightforward to solve but computationally difficult. Often, however, with a little thought we can guess the form of the answer and then use differentiation to find the antiderivative. In this problem, you are going to learn the technique itself in one special case and then generalize it to other situations.

a. Let A and B be real numbers and consider the integral $\int \cos(Ax)e^{Bx}\,dx$. From the integrals of $\cos(x)$ and e^x, we might guess the antiderivative is a multiple of $\sin(Ax)e^{Bx} + C$. Show this is not correct. By considering the derivative of $\cos(Ax)e^{Bx}$, we might guess that an antiderivative involves a linear combination of trigonometric functions times e^{Bx}. Use this idea to find the antiderivative of $\cos(Ax)e^{Bx}$ and use this to evaluate

$$\int \cos(3x)e^{4x}\,dx.$$

b. Using the technique of part (a), derive and explain a technique for evaluating $\int \cos(Ax)\sin(Bx)\,dx$, where $A^2 \neq B^2$, without performing any integration.

c. Many books have the following formula:

$$\int \cos(Ax)\sin(Bx)\,dx = \frac{-\cos(B-A)x}{2(B-A)} + \frac{-\cos(B+A)x}{2(B+A)} + C.$$

Show that this is equal to your answer to part (b).

d. Let n be a positive integer and $P(x)$ a polynomial of degree n. Use the technique above to derive a technique for evaluating $\int P(x)e^{Ax}\,dx$. Explain your technique in detail. Evaluate the following integral:

$$\int (4x^6 - 2x^5 + 2x^3 - 10x^2 + 5)e^{4x}\,dx.$$

Extra credit: Can you think of other products of classes of functions where we could use these ideas? Discuss the technique in these cases.

Comments

Requires: integration by parts, trigonometric identities

Teaches: techniques of integration, undetermined coefficients

The goals of this project are to teach students the relationship between differentiation and integration and how the simpler operation of differentiation can be used to perform the harder operation of integration.

In its original version, the students did not understand the instructions on their own. The project has been extensively rewritten, but has not been reassigned in this form.

The project develops a technique that in many instances can be used in place of integration by parts. Instructors may want to talk about this with students. Instructors may have to tell students about linear combinations.

Fourier Polynomials

Let f be a continuous function on the interval $[-\pi, \pi]$. We define the Fourier coefficients of f by the formulae:

$$a_n = \frac{1}{\pi} \int_{-\pi}^{\pi} f(x) \cos(nx)\, dx \quad \text{and} \quad b_n = \frac{1}{\pi} \int_{-\pi}^{\pi} f(x) \sin(nx)\, dx, \quad n = 0, 1, 2, \ldots .$$

For $N = 1, 2, \ldots$, we define the "Fourier polynomials" of f to be

$$P_N(x) = \frac{a_0}{2} + \sum_{n=1}^{N} [a_n \cos(nx) + b_n \sin(nx)].$$

Fourier coefficients and polynomials are named for the French mathematician Joseph Fourier. In a speech for the French Academy of Sciences in 1807, Fourier proposed that these polynomials could be used to approximate arbitrary functions on the interval $[-\pi, \pi]$. He was led to study these polynomials and make his bold statement by studying heat flow.

Unfortunately, the great French mathematicians of the time did not agree and did not take his ideas seriously. They were wrong. Today, the ideas introduced by Fourier are used to study phenomena that exhibit wavelike (*periodic*) behavior, such as sound and light. These ideas have been used in physics, engineering, and even economics. This problem is an introduction to some of Fourier's work.

a. Let $f(x) = x$. Find all of the Fourier coefficients of f. What are the first five Fourier polynomials P_N associated to f? Plot the function f and these five polynomials.

b. Let f be an odd function, continuous on the interval $[-c, c]$. Prove that

$$\int_{-c}^{c} f(x)\, dx = 0.$$

c. In part (a), something special happened because the function, $f(x) = x$, is an odd function on $[-\pi, \pi]$. What pattern did you notice? Next, let f be any odd function on $[-\pi, \pi]$. Make a conjecture about the Fourier coefficients of f. Prove your conjecture.

d. Another special collection of functions on an interval $[-c, c]$ is the even functions. After thinking about what happens for the odd functions, make a conjecture about the Fourier coefficients of an even function. Let $f(x) = x^2$; check your conjecture using this even function. Explain why your conjecture is true for any even function.

Extra credit: Why did Fourier think this would be a good technique for approximating periodic functions?

Comments

Requires: integration by parts, summation notation

Teaches: application of integration

Students seemed to like this project and did reasonably well on it. They were successful at solving the extra credit. They had a hard time seeing that they could apply part (b) to parts (c) and (d).

Simpson's Rule

In this problem, you will derive Simpson's rule for estimating

$$\int_a^b f(x)\, dx.$$

a. To begin, you will get an estimate for

$$\int_{-1}^1 \sqrt{1 + x^3}\, dx.$$

You will need to use the fact that given three points in the plane with distinct x-coordinates, there is a polynomial of degree at most 2 whose graph passes through all three points. Let $f(x) = (1 + x^3)^{1/2}$. Write the equation of such a polynomial, $y = p(x)$, that passes through the three points $(-1, f(-1))$, $(0, f(0))$, and $(1, f(1))$. Graph f and p together over $[-1, 1]$.

b. Next, compute

$$\int_{-1}^1 p(x)\, dx$$

and show that this is Simpson's estimate for

$$\int_{-1}^1 \sqrt{1 + x^3}\, dx$$

with two subintervals.

c. Let f be a continuous function and h a fixed positive number. Find a formula for the quadratic function p whose graph passes through the points $(-h, f(-h))$, $(0, f(0))$, and $(h, f(h))$. Show that

$$\int_{-h}^h p(x)\, dx = \frac{h}{3}[f(-h) + 4f(0) + f(h)].$$

d. Let c be a fixed number and let p be the quadratic function whose graph passes through the points $(c - h, f(c - h))$, $(c, f(c))$, and $(c + h, f(c + h))$. Find

$$\int_{c-h}^{c+h} p(x)\, dx$$

in terms of $f(c - h)$, $f(c)$, and $f(c + h)$.

e. You are ready to derive Simpson's rule. Let k be an integer and divide $[a, b]$ into $2k$ sections. Group them into k pairs of adjacent sections. Over each pair, estimate the integral of f as in part (d). Show that when these k separate estimates are added, Simpson's estimate results.

Comments

Requires: integration

Teaches: Simpson's rule

This problem has never been assigned. It seems computationally and symbolically difficult. A proof of Simpson's rule can be found in the body or outlined in the problem sections of many books.

Integration of Some Special Rational Functions

You are going to derive a method for integrating rational functions of the hyperbolic functions. You will have to learn about the hyperbolic functions and generalize a technique used for the trigonometric functions.

a. You will first learn to integrate any rational function of $\cos\theta$ and $\sin\theta$. The main idea is to use a clever (and intricate) substitution to reduce rational functions of $\cos\theta$ and $\sin\theta$ to rational functions of u. Read about integrating rational functions of $\sin\theta$ and $\cos\theta$. Using that technique, compute

$$\int \frac{d\theta}{4\cos\theta + 5\sin\theta}.$$

b. Next you will need to learn about the hyperbolic functions. Look for similarities between the hyperbolic and trigonometric functions. What are the identities for hyperbolic functions analogous to

$$\sin^2\theta + \cos^2\theta = 1 \quad \text{and} \quad \tan^2\theta + 1 = \sec^2\theta?$$

   Derive these analogous identities.

c. Show that $\frac{d}{dt}\tanh t = \operatorname{sech}^2 t$.

d. Your goal is to find a method for integrating rational functions of $\cosh t$ and $\sinh t$. Look back at the substitution used with the trigonometric functions. Make the analogous substitution for u in terms of hyperbolic functions. Find formulae for $\sinh t$, $\cosh t$, and dt in terms of u and du. With these formulae, you can change any rational function of $\cosh t$ and $\sinh t$ into a rational function of u.

e. Use the technique of part (d) to find the following integrals:

$$\int \frac{1}{\cosh t + \sinh t}\, dt \quad \text{and} \quad \int \frac{2e^t}{2 + e^t + e^{-t}}\, dt.$$

Extra credit: There are other places where we could use hyperbolic functions instead of trigonometric functions. Derive some hyperbolic function versions of the standard substitutions using trigonometric functions. Using several examples, compare the ease of use of hyperbolic to trigonometric substitution.

Comments

Requires: techniques of integration

Teaches: hyperbolic functions

The students seemed to like this problem and to feel that they were learning some real mathematics. The material on rational functions of the trigonometric functions was not covered in the course, so the students had to learn that on their own.

Instructors should be aware that many students will have a hard time evaluating the second integral in part (e).

Students without a good grounding in the trigonometric functions will have a hard time with this problem. This is probably good, since the project gives them a chance to review trigonometry.

Life on the Hyperbolic Curve

We have studied the use of trigonometric substitution in finding the antiderivatives of certain special types of functions. The success of these substitutions depends on certain trigonometric identities and the integrals of special combinations of trigonometric functions. In this project, you will learn about hyperbolic functions and how such functions may be used in place of the trigonometric functions in the substitution method for finding antiderivatives.

First, read about the hyperbolic functions. Any good calculus book can serve as a reference. Then, you will be ready to find the antiderivatives of the functions given below. You are to use substitutions with hyperbolic functions to find the antiderivatives and you must justify any identities or integration formulas that you use. Perhaps you might put all the justifications in an appendix and give references to the results from the appendix as you use them. Since you are using the hyperbolic functions only as a tool to find the antiderivatives, your answers should involve neither hyperbolic functions nor their inverses.

Find an antiderivative of each of the functions given below.

a. $(9x^2 + 4)^{-3/2}$

b. $\sqrt{x^2 - 4}$

c. $\frac{1}{x^2\sqrt{16-x^2}}$

d. $\frac{1}{\sqrt{1-x^2}}$

In part (d), you probably expected to get $\arcsin(x) + C$. How do you reconcile this with your answer?

Comments

Requires: exponential functions, techniques of integration

Teaches: hyperbolic functions, techniques of integration

This project was assigned in an accelerated calculus course. The hyperbolic functions had not been discussed in class, so some reading was required. One of the students was a high school teacher who commented that her students ask her about the hyperbolic functions and this project gave her an understanding of those functions.

The instructor will have to decide what constitutes a correct answer for the last paragraph.

A Special Technique for Integration

Often, two integrands can look similar while one is easy to integrate and the other very hard. Consider, for example, the integrals:

$$\int_0^{\pi/2} \frac{\sin x}{\sin x + \cos x}\, dx, \quad \int_0^{\pi/2} \frac{\sin^2 x}{\sin^2 x + \cos^2 x}\, dx,$$

and

$$\int_0^{\pi/2} \frac{\sin^{10} x}{\sin^{10} x + \cos^{10} x}\, dx.$$

The second one is easy to integrate while the other two are not. Still, you are going to evaluate

$$\int_0^{\pi/2} \frac{\sin^n x}{\sin^n x + \cos^n x}\, dx$$

where n is any positive integer. To do this, you will need one trick and some ingenuity. We will supply the trick.

Before you work on the entire problem, evaluate the second integral $(n = 2)$ without doing anything "hard." Next, read about integrating rational functions of $\sin x$ and $\cos x$ and use that idea to evaluate the first integral $(n = 1)$.

We could use this approach for larger values of n, but imagine how hard it would be. We need to find an easier way.

Now, here is the trick. Let f be a continuous function over the closed interval $[0, a]$. Then,

$$\int_0^a f(x)\, dx = \int_0^a f(a - x)\, dx$$

Prove that this equality is true.

Use this result to evaluate

$$\int_0^{\pi/2} \frac{\sin^n x}{\sin^n x + \cos^n x}\, dx$$

for any value of n. Remember, you will need to use some ingenuity here also. Make sure that your answer agrees with your previous results for $n = 1$ and $n = 2$.

Comments

Requires: techniques of integration, trigonometric identities

Teaches: change of variables

This is an interesting problem which has the students learn a technique of integration on their own (about rational functions of the trigonometric functions) and gives them an idea about why theorems can be important. Students who did not mind doing a purely theoretical problem enjoyed this one.

Cubics and Coordinate Systems

In this project you will discover some surprising facts about the graphs of cubic polynomials.

a. Carefully graph the polynomial function $f(x) = x^3 - 3x^2$. Be sure to find and identify critical points and inflection points.

b. By now you have discovered that f has a relative maximum at $x = a$ and a relative minimum at $x = b$. Let $P = (a, f(a))$ and $Q = (b, f(b))$. Draw the line through P and Q and let $y = L(x)$ be the equation of that line. Evaluate:

$$\int_a^b [f(x) - L(x)]\, dx.$$

c. Repeat parts (a) and (b) above for other cubic polynomials of your choosing until you see a pattern. Make a conjecture based on that pattern.

d. Return now to the function f given in part (a). The chord PQ and the graph of f intersect in three points, P, Q, and a third point R. Two regions are thus formed. Find the area of each. What do you notice? Now explain the results observed in parts (b) and (c) in terms of what you have discovered.

e. Read about translations of axes and symmetry. Begin with the graph of $y = x^3 - 3x^2$ and translate the axes in such a way that the new origin is at the point R you found in part (d). What is the equation of the curve in the new coordinate system? Use this equation and symmetry to further explain your discoveries.

f. Make a conjecture concerning the symmetry of the graph of a general cubic polynomial function $f(x) = ax^3 + bx^2 + cx + d$. Now you are to prove that your conjecture is true. You will probably want to use translation of axes as you did in part (e).

Comments

Requires: critical points, inflection points, area, integration, symmetry

Teaches: translation of axes, symmetry

Students perceived this project as being easy and consequently stumbled when they got to part (f). For that reason part (f) has been rewritten to make the instructions more explicit. However, it is probably worthwhile to remind your students what is expected as a proof.

Sticking the Tank

You have just been hired by the Environmental Protection Agency (EPA) under the Superfund program to measure the level of toxic wastes in buried tanks across New Mexico. Most of these tanks are cylinders, with their axes horizontal. You are to "stick the tank" by inserting a stick through a hole in the center of the top until it touches the bottom, then pulling it out and reading off the liquid level showing on the stick. They have hired you because you know calculus; they have faith that you can convert the "height on the stick" reading to "filled volume in the tank."

Assuming that the cross-sectional radius of the tank is R and its length is L, calibrate the stick for them. That is, convert height showing on the stick to volume of liquid. Check your results by doing the calculation in two separate ways:

a. Evaluate a definite integral that gives the filled volume in terms of the height h on the stick. DO NOT use tables! (Suggestion: place the origin of your coordinate system at the center of the circular cross-section. Make a sketch!)

b. Use elementary geometry and trigonometry (no calculus) to obtain the volume.

c. Show that your results for (a) and (b) are equal.

Since you have been so successful in such endeavors, the EPA sends you out to stick a tank that has the shape of an elliptical cylinder, i.e., whose cross section is an ellipse instead of a circle. The major axis of the ellipse is horizontal. Calibrate the stick for them, using calculus.

Extra credit: Calibrate the stick for tilted tanks of circular and elliptical cross sections.

Comments

Requires: volume by slices, trigonometric substitution

Teaches: geometry of trigonometric substitutions.

Students who chose an unfortunate substitution had a hard time recognizing the equality of the answers in parts (a) and (b). You may direct them to use the identity $\cos^{-1} x = \frac{\pi}{2} - \sin^{-1} x$. Also, it is more convenient to place the origin at the center of a circular slice.

The Barge Captain's Wager

You are a barge captain on the Erie canal. It is payday, and in a dockside establishment you make a $100 bet with Big Jim, the mule-skinner who hauls the barge. He claims (boisterously) that, if the barge starts with the tow line at right angles to the shore, then the barge will move in a straight line as he drives his mule in a straight line along the shore. Prove him wrong and win (not $100 but a good grade in calculus) by following the steps below. Assume that, whatever curve the barge follows, the tow line will always be tangent to this curve. Assume that the tow line has a constant length of 130 feet. Assume that the mule walks along the y-axis.

a. Find an expression for the slope of the tangent line to the curve, which gives $\frac{dy}{dx}$ as a function of x.

b. Use your result from part (a) to show that this curve is not a straight line and collect your wager.

c. Find a function, whose derivative you found above, that matches the initial position of the barge at the start of the tow and thus find the path of the barge. DO NOT use tables!

Nellie can't help but overhear your conversation with Big Jim, and decides to get in on the action. She's the type, however, who only goes with a winner. She says, "The rope on my old rowboat is exactly 50 feet long. Now I'm going to row up the canal right next to the tow-path for whatever distance you tell me and tie up. Then I'll row out 50 feet perpendicular to the shore and wait for your barge to come by. If it looks like it's coming straight for me, I'll jump aboard and go on to Buffalo with you—provided you can tell me exactly how fast the barge will be going when we meet. Otherwise, you just keep going and I go back with Big Jim. Do we have a deal?"

d. Assuming the mule is walking at four miles per hour, tell Nellie exactly how far up the tow path to tether her rowboat and how fast you will be going when you meet, so she can jump aboard.

Hint: The speed of your barge is given by the expression

$$\sqrt{\left(\frac{dx}{dt}\right)^2 + \left(\frac{dy}{dt}\right)^2}$$

when its position is given by $(x(t), y(t))$.

Comments

Requires: tangent line, integration of trigonometric functions

Teaches: speed, differential equations

The students enjoyed this problem. They often made up sidelines to the story on their own.

Space Capsule Design

You are part of a team of engineers designing the Apollo space-capsule. The capsule is composed of two parts:

1. A cone with a height of 4 meters and a base of radius 3 meters;

2. A re-entry shield in the shape of a parabola revolved about the axis of the cone, which is attached to the cone along the edge of the base of the cone. Its vertex is a distance D below the base of the cone.

   Assume the capsule has uniform density ρ. Your project director has specified that the center of mass of the capsule should be below the center of mass of the displaced water because he believes this will give the capsule better stability in heavy seas. He has given your team the task of finding values of the design parameters D and ρ so that the capsule will float with the vertex of the cone pointing up and with the waterline 2 meters below the top of the cone, in order to keep the exit port $\frac{1}{3}$ meter above water.

   a. Show your project director that this task is impossible; i.e., there are no values of D and ρ that satisfy the design specifications.

   b. Prove that you can solve this dilemma by incorporating a flotation collar in the shape of a torus (doughnut). The collar will be made by taking hollow plastic tubing with a circular cross section of radius 1 meter and wrapping it in a circular ring about the capsule, so that it fits snugly. The collar is designed to float just submerged with its top tangent to the surface of the water. Show that this flotation collar makes the capsule plus collar assembly satisfy the design specifications. Find the density ρ needed to make the capsule float at the 2 meter mark. Assume the weight of the tubing is negligible compared to the weight of the capsule, that the design parameter D is equal to 1 meter, and the density of water is 1.

   Your investigations will be guided by a physical principle (I), a formula, which you will derive (II), and a theorem which you will state and prove (III).

   I. *Archimedes' Principle*: A body floats in a fluid at the level at which the weight of the displaced fluid equals the weight of the body.

   II. Consider the region bounded by the graph of $x = f(y)$, the y-axis, and the lines $y = c$ and $y = d$. Derive a formula for the center of mass of the solid formed by revolving this region about the y-axis. Explain why the center of mass is on the y-axis so you just have to find $\bar{y}$.

III. Suppose a body is made up of two pieces, which are solids of revolution about the y-axis. Conjecture and prove a result relating the center of mass of the composite body to the centers of mass of the pieces.

Hint: Suppose one piece has center of mass at $\bar{y}_1$ and mass m_1 and the other has center of mass at $\bar{y}_2$ and mass m_2. Calculate the center of mass of the composite object by assuming m_1 is concentrated at $\bar{y}_1$ and m_2 is concentrated at $\bar{y}_2$. The formula you obtain for $\bar{y}$ should be your conjecture. Proceed with the proof using the definitions of $\bar{y}_1$, $\bar{y}_2$, and $\bar{y}$ in terms of integrals.

Comments

Requires: center of mass, volume by slices

Teaches: center of mass

The students worked on this project in groups. It was assigned in a one-variable calculus course, so care had to be taken to use circular symmetry. The students had fun with this project; some wrote up their results as a NASA technical report! Students seemed to grasp the point in part (III) more quickly if the axis was turned on its side and the masses presented as balancing on a "teeter-totter."

Comparison of Improper Integrals

Let f and g be continuous nonnegative functions on the infinite interval $[a, \infty)$. In this problem, you will consider the convergence of improper integrals and how inequalities involving f and g can be used to give information about the convergence of the integrals. Before continuing, review the material on limits at infinity since this project relies on that idea.

a. Suppose that we have the inequality $0 \leq f(x) \leq g(x)$. We have seen that if $\int_a^\infty g(x)\,dx$ converges then $\int_a^\infty f(x)\,dx$ converges.

    i. What can you say about $\int_a^\infty f(x)\,dx$ if $\int_a^\infty g(x)\,dx$ diverges?

    ii. What can you say about $\int_a^\infty g(x)\,dx$ if $\int_a^\infty f(x)\,dx$ diverges?

    iii. What can you say about $\int_a^\infty g(x)\,dx$ if $\int_a^\infty f(x)\,dx$ converges?

    Give examples to support your answers.

b. Let $k > a$ and $M > 0$. Prove the following two statements.

    i. $\int_a^\infty f(x)\,dx$ converges if and only if $\int_k^\infty f(x)\,dx$ converges.

    ii. $\int_a^\infty f(x)\,dx$ converges if and only if $\int_a^\infty Mf(x)\,dx$ converges

c. Set $L = \lim_{x\to\infty} \frac{f(x)}{g(x)}$. Suppose that $0 < L < \infty$. Discuss the relationship between the convergence of $\int_a^\infty f(x)\,dx$ and $\int_a^\infty g(x)\,dx$ as in part (a), and state a conjecture. Prove your conjecture.

d. What can you say if $L = 0$? Use this to prove that $\int_1^\infty e^{-2x}x^t\,dx$ converges if $t \geq 0$.

Comments

Requires: limits at infinity, improper integrals

Teaches: limits, comparison of integrals

    We had hoped that this project would tie a lot of ideas together for the students. Since it was assigned the semester before we cover series, we also hoped it would help prepare students for that topic. Unfortunately, students had a very hard time with this problem. The difficulty related to their lack of understanding of limits; that is why a comment has been added to the end of the introductory paragraph telling them to review limits. This project might be more successful if it is assigned to groups of two or three students.

It may be difficult for students to discover the function used in the comparison for the integral in part (d).

Evaluation of Special Improper Integrals

This problem is concerned with evaluating some improper integrals. In particular you will use an improper integral over an interval of infinite length to evaluate an integral of a function not defined at one end point. This will involve a special function Γ which arises in many applications in the sciences.

a. Evaluate $\int_0^1 x(\log x)^2 \, dx$.

b. Explain how you would evaluate $\int_0^1 x^8(\log x)^7 \, dx$, but do not actually compute it. Would your method work if the exponent '8' were replaced by '$8\frac{1}{2}$', or if '7' were replaced by '$7\frac{1}{2}$'? Why?

c. Define the function $\Gamma(t) = \int_0^\infty e^{-x}x^{t-1} \, dx$ for
   $t > 0$.

   i. Calculate $\Gamma(1)$.
   ii. Let t be a fixed number, $0 < t < 1$. Explain why $\Gamma(t)$ is an improper integral. Then, show that the integral that defines $\Gamma(t)$ is convergent.
   iii. For $t > 0$, show that $\Gamma(t+1) = t\Gamma(t)$.

d. Let n be a positive integer. Use part (c(iii)) to find an equation relating $\Gamma(n)$ and $n!$. Read about mathematical induction and prove your equation is correct.

e. Let m be a positive real number and let n be a positive integer. Find an equation relating $\int_0^1 x^m(\log x)^n \, dx$ to Γ and use this to find a formula for this integral in terms of m and n.

f. Use part (e) to evaluate three integrals in (a) and (b).

Comments

Requires: improper integrals, integration by parts

Teaches: factorial, mathematical induction, gamma function

Since calculus books do not tend to have a discussion of induction, it may be a good idea to give students a reference for part (d). This problem was originally assigned asking the students only to find the relationship in part (d) but not prove it. They had no way to be sure their relationships were correct. Having them prove the relationship is also a good way to get students in the calculus to use induction.

Although the integral in part (a) is not an improper integral, most students will not recognize that and will treat it as an improper integral.

The Ferris Wheel

You are standing on the ground at a distance of 25 meters from the bottom of a ferris wheel on which your friend is riding. You and the ferris wheel are in the same plane. Your arm is at the same level as the bottom of the wheel. The ferris wheel has a radius of 7 meters and makes one revolution every 12 seconds. The bottom of the ferris wheel is moving towards you. You throw a ball at 20 meters/second at an angle of 60° above the horizon, releasing it at the instant that she is halfway up the ferris wheel on your side. Your friend will catch the ball if it comes within one third of a meter of her. Does she catch it? Assume that the acceleration due to gravity equals −9.8 meters/second^2 and neglect air resistance.

a. Find the distance between the ball and your friend as a function of time.

b. Attempt to minimize the distance. What happens when you try to minimize this function? To simplify the computations, you might minimize the square of the distance. Explain why this leads to the same answer.

c. Read about Newton's method. Apply Newton's method to estimate how close the ball gets to your friend. You may assume that your estimate is sufficiently accurate if the estimate you get changes by less than 0.1 meter between two consecutive iterations. Be sure that you understand when you can consider your answer to be sufficiently accurate. Show and explain your work.

d. Determine whether or not your friend caught the ball.

Extra credit: Determine the effect of changing the speed or the angle at which you throw the ball.

Comments

Requires: parametric equations

Teaches: Newton's method, parametric equations

This problem relates parabolic motion to circular motion and brings together many ideas (some new to the students) in a physically reasonable setting. Students had to teach themselves Newton's method in order to solve this one. In part (c), the students are asked to run Newton's method until the distance changes by less than 0.1 meter. Several students stopped the process when the time changed by less than 0.1 second. A comment has been added to part (c) to help avoid this problem.

Although there is no mention of computers, this is a very good problem to solve using computers, and several students did so. One must work out parts (a), (b), and the beginning of (c) before turning to the computer to do the iterations in Newton's method. On the other hand, with any reasonable choice of a starting time, one can stop after two or three iterations, so students can do this project using only a calculator.

The All-Purpose Calculus Project

Carol, a local bicycle racing star, is moving at a constant velocity k meters per second, as she passes a refreshment station in the shape of a giant laboratory flask with cross sectional radius, measured in feet, given as a function of height z from the ground by the formula

$$r(z) = \frac{10}{\sqrt{z}}.$$

Sand is flowing from the flask at a constant rate into a parabolic mirror along a line parallel to light rays from the star Alpha Centauri. The parabolic mirror is located inside a greenhouse with the south wall replaced by a huge sloped piece of glass (which is expensive) that was purchased at a store located along an existing road at a point at which the shortest route connecting two towns not on the road intersects the road.

An observer watches Carol from a ferris wheel of radius 7 meters that makes one revolution per minute and is floating on a barge, towed by a mule walking a straight line along the shore of the Erie Canal.

a. Use the precise definition of limit to find the center of gravity of the barge.

b. Can you think of other shapes for the laboratory flask that will simplify the problem?

c. Write the equation of the tangent line and discuss your solution.

d. If Alpha Centauri collapses to a black hole, what will be the position of the observer on the ferris wheel?

e. If the light rays from Alpha Centauri and the rope between the mule and the barge form a triangle of constant area, use mathematical induction to show that the barge will float with the exit port $1/3$ meter above the water.

Extra credit: How would your solution change if the greenhouse were replaced by a pup tent made in the shape of a dome with volume 2.2m^3, of minimum cost, where the material for the floor of the tent costs 1.4 times as much as the rest?

Comments

Requires: independent reading, second fundamental theorem of calculus, arbitrary constants, finite group theory, intuitive arguments

Teaches: proof, mathematical induction, counterintuitive results, applying theorems, direction cosines, generalization, geometric intuition, precise definition of limit

If you are pressed for time and can only assign one project, this is the one to use. This project may be found in some calculus textbooks or solution manuals. Preferably, this should be a group project with group sizes varying from 3 to 16. Be sure to warn students that they must justify everything. Some students tried to use Newton's method, but applied it incorrectly. This project has been rewritten extensively. A more theoretical version without the story line has been given before. Instructors may have to talk to the students about linear combinations. Students liked this project because it was applied. Many made up their own story line. It has never been assigned. If you wish to create some variations on this project, you might note that the underlying ideas are directly related to Fermat's last theorem. When the projects are turned in, you might want to spend a bit of class time discussing how the results are related to the number of primes between $\log n$ and n.

Be careful if you try different shapes with the extra credit part. They vary enormously in size.

This may be a one year project.

Estimating the Derivative

This project concerns two different ways of approximating the derivative of f at $x = a$, i.e., two different "estimates" of $f'(a)$. Your job is to discover which is more accurate, and then to discover why. Assume that f is three times continuously differentiable.

The first estimate of $f'(a)$ involves the familiar difference quotient:

$$f'(a) = \frac{f(a + \Delta x) - f(a)}{\Delta x} + \epsilon_1(\Delta x), \tag{1}$$

where $\epsilon_1(\Delta x)$ is the error in this estimate and depends on Δx.

a. Choose the function $P(x) = x^3 - 3x$ as an experimental subject. Sketch its graph. Choose two values of a where the graph is concave upward and two values where the graph is concave downward, and evaluate ϵ_1 for several different values of Δx at each a. Do you notice a pattern? How does ϵ_1 vary with Δx? With $P''(a)$?

b. Now consider a general function f whose graph is concave upward in the region between a and $a + \Delta x$. Sketch a picture showing the numerator and denominator in the difference quotient, the tangent line at $x = a$, and the error ϵ_1, all superimposed on the graph of f. Argue from your sketch that the sign and magnitude of the error should depend upon the signs and magnitudes of both Δx and $f''(c_1)$, for some c_1 between a and $a + \Delta x$.

c. How do you expect ϵ_1 to vary with Δx? With $f''(c_1)$? What do you expect to happen to ϵ_1 as $\Delta x \to 0$?

   Hint: Think about the definition of the derivative.

d. From now on assume $\Delta x > 0$. Using a theorem from calculus, show that the error is given by

$$\epsilon_1 = -\left(\frac{f''(c_1)}{2}\right)\Delta x, \quad a < c_1 < a + \Delta x.$$

   Find c_1 for $P(x) = x^3 - 3x$, $a = \frac{1}{2}$, and $\Delta x = \frac{1}{4}$.

   The second estimate involves the central difference quotient:

$$f'(a) = \frac{f(a + \Delta x) - f(a - \Delta x)}{2\Delta x} + \epsilon_2(\Delta x), \tag{2}$$

   where $\epsilon_2(\Delta x)$ is the error in this estimate.

e. Evaluate the error for the same function P and the same values of a and Δx you used in part (a). Are the errors ϵ_2 greater or smaller in absolute value than the corresponding errors ϵ_1 for your first estimate?

f. Let the graph of f be concave up between a and $a + \Delta x$. Sketch pictures showing the numerator and denominator of the central difference quotient in formula (2), the tangent line at $x = a$, and ϵ_2 superimposed on the graph of f. Make a similar sketch for an f whose graph is concave down. Which difference quotient, the one appearing in (1) or (2), do you expect to be a better estimate of the derivative for small values of Δx? Argue on the basis of these pictures and those in part (b).

g. Show that

$$\epsilon_2 = -\left(\frac{f'''(c_2)}{6}\right)(\Delta x)^2, \quad a - \Delta x < c_2 < a + \Delta x.$$

Suppose that a is not an inflection point. Which estimate would you expect to have an error with smaller absolute value as $\Delta x \to 0$?

h. Compare the analytical result of part (g) with your pictures in parts (b) and (f). What happened to the error caused by the second derivative in part (d) when we went to the central difference estimate, whose error was given in part (g)?

Comments

Requires: Taylor's theorem, derivative form of Taylor remainder

Teaches: Taylor's theorem, error estimation

This project has been extensively revised since it was first used in order to give students a graphical appreciation of the error and its relation to the second derivative.

The Deft Fly

Two bicycles speed straight towards each other from 60 miles apart, each moving at 10 miles per hour. A fly (whose speed is 20 miles per hour), initially perched on one of the bicycles, begins flying back and forth between the bicycles just as they start. The deft fly makes instant turnarounds with no change in speed. As the bicycles crash, the fly is squashed. How far did the deft fly travel? Solve this problem two ways.

First use an infinite series. For this you will find yourself needing a formula for the distance the fly travels on its nth trip between bicycles. After some initial calculations, you may be able to guess the correct formula. To prove that your guess is correct, you will need to use "mathematical induction." Read about mathematical induction and study some examples of its use to help you learn how to apply it here.

Now solve this problem using a completely different argument that a twelve-year-old could understand.

Comments

Requires: geometric series, mathematical induction

Teaches: mathematical induction, infinite series

Students may need help in properly applying mathematical induction.

A Circumstance in which f is Analytic

We say a function f is *analytic* at a point a if f equals its Taylor series at a in an open interval that contains a. Suppose f is infinitely differentiable. Assume also that there is a fixed positive integer m such that $|f^{(n)}(x)| \leq n^m$ for all x, provided n is sufficiently large. Show that for any value of a,

$$f(x) = \sum_{k=0}^{\infty} \frac{f^{(k)}(a)}{k!} (x-a)^k \qquad \text{for all } x.$$

At what points is f analytic? Explain.

Hint: You may wish to read about the limit $\lim_{n \to \infty} \frac{r^n}{n!}$ (r constant).

Comments

Requires: Taylor series, Taylor's theorem

Teaches: Taylor's theorem, factorial, limits

The hint has been added since students generally needed to be pointed in this direction anyway.

The Number e is Irrational

Fill in all the missing steps in this sketch of a proof that e is irrational. The proof is by contradiction.

Proof.　Suppose to the contrary that e is rational. Then $e = \frac{p}{q}$ where p and q are positive integers. Let

$$M = q! \left\{ e - \sum_{k=0}^{q} \frac{1}{k!} \right\}.$$

Explain why M is a positive integer. Explain why

$$M = q! \sum_{k=q+1}^{\infty} \frac{1}{k!} = \sum_{k=q+1}^{\infty} \frac{q!}{k!}.$$

Compare this favorably with a geometric series to show that $M < \frac{1}{q}$. Explain exactly how this completes the proof that e is irrational.

Comments

Requires:　geometric series, comparison test

Teaches:　manipulating infinite series, proof by contradiction

Students may need references on proofs by contradiction.

Agility with Power Series

If $f(x) = \sum_{n=1}^{\infty} nx^n$, calculate $(1-x)f(x)$ directly and use the result to find a closed formula (i.e., no infinite sum) for $f(x)$. Then, use a similar idea to find a closed formula for $g(x) = \sum_{n=1}^{\infty} n(n+1)x^n$.

You will need some facility working with indexing in sums. Here is an example to help. Suppose we wish to add $\sum_{n=1}^{\infty} a_n x^n$ to $\sum_{n=1}^{\infty} b_n x^{n+1}$. To add them term-by-term to create a single power series, we need the powers of x to agree in the general terms. We arrange this by first rewriting $\sum_{n=1}^{\infty} a_n x^n$ as $\sum_{m=1}^{\infty} a_m x^m$ (to avoid confusion between the different uses of n). Now in $\sum_{n=1}^{\infty} b_n x^{n+1}$ we change the notation ("reindex") by letting $n+1 = m$, which is the same as $n = m - 1$. So now we have $\sum_{n=1}^{\infty} b_n x^{n+1} = \sum_{m=2}^{\infty} b_{m-1} x^m$. Notice the new sum starts at $m = 2$. Why? Now we add:

$$\sum_{n=1}^{\infty} a_n x^n + \sum_{m=2}^{\infty} b_{m-1} x^m = a_1 x + \sum_{m=2}^{\infty} a_m x^m + \sum_{m=2}^{\infty} b_{m-1} x^m$$

$$= a_1 x + \sum_{m=2}^{\infty} (a_m + b_{m-1}) x^m.$$

We obtain a single power series.

Comments

Requires: power series, differentiation of power series, geometric series

Teaches: power series, index notation, reindexing

The example is included since students had a hard time with reindexing.

Speedy Series for Lethargic Logarithms

Many handbooks of mathematics list several series for the natural logarithm. Find such a book and peruse these series. Typically, very little information is given either to justify these series or to help the user decide which might be appropriate for the task at hand. Put yourself in the position of someone who needs a series to compute logarithms over a wide range of values. In this project you will develop useful information about two series for the natural logarithm.

a. Find the Taylor series of $\log x$ expanded about 1. What is its radius of convergence? Compute all the partial sums, for up to ten terms, of $\log 0.5$ and $\log 2$. How many terms would be needed to have ten decimal places of accuracy when $x = 0.1, 0.5, 1, 2$? How might you use this series to compute $\log 10$?

b. To get a more rapidly converging series, consider the transformation of variables,

$$x = \frac{y+1}{y-1}.$$

Develop $\log x$ as a Taylor series in y about 0. What is its radius of convergence in terms of y? What is the range of convergence in terms of x? Compute all the partial sums, for up to ten terms, of $\log 0.5$ and $\log 2$. How many terms would be needed to obtain ten decimal places of accuracy when $x = 0.1, 0.5, 1, 2, 10$?

c. Compare your results for parts (a) and (b) by constructing appropriate tables that summarize your calculations.

Extra credit: Write the inverse transformation of variables as a function, $y = t(x)$. Prove that

$$t\left(\frac{1}{x}\right) = -t(x).$$

How might this relationship be used to avoid computing the series in (b) for negative y? Is there any value in avoiding such y? Compare with the analogous situation for the series in (a).

Comments

Requires: Taylor series

Teaches: Taylor series, rate of convergence

This project has never been assigned.

Most students have considerable difficulty in understanding series and appreciating their value. This project was designed to alleviate this by having students compute the logarithm for two very different series and estimate the number of terms needed for high accuracy. Correlating these calculations with the theory should clear up some of their misconceptions.

Erroneous Exponentials

Translating Taylor series into useful computer programs for evaluating transcendental functions is not straightforward; there are many pitfalls. The object of this project is to fall into some of these. Some of these pitfalls are due to misapplications of the theory and others are due to peculiarities of calculators.

a. Write a program on your calculator to compute the partial sums of the Maclaurin series for e^x. Do this in a straightforward way by computing the powers, dividing each by the appropriate denominator and successively summing the terms. Truncate the series, that is, break the loop, when the next term in absolute value is less than a parameter ϵ. Compute e^2 and e^{-2} with this program when $\epsilon = 0.1$. Compare these results with what you get using the exponential button on your calculator. Why is one answer correct to within ϵ and the other not?

b. Compute e^9 with $\epsilon = 10$. (Although this ϵ looks ridiculously large, it is small compared to e^9.) What went wrong? Modify your algorithm so that the answers in both parts (a) and (b) will be accurate to within ϵ, as follows. A stopping rule can be obtained by bounding the remainder—the difference between the Taylor polynomial and the Taylor series—by a geometric series, whose value can be expressed as a rational function. Be sure to take into account that a geometric series has a finite radius of convergence. Why won't the derivative form of the remainder help here?

c. Computing x^n and then dividing by $n!$ for each term is bad numerical practice since one may exceed the capacity of the machine and it takes time. Modify your program again by computing each new term iteratively in terms of the previous term. This should make the program run faster.

d. For the last straw, compute e^{-50} with $\epsilon = 0.1$, and explain what went awry. Using part (c), how can you modify the algorithm to get around this?

Comments

Requires: programmable calculator, Taylor series

Teaches: geometric series, numerical methods

This project has never been assigned.

It is roughly the numerical analog of *Speedy series for lethargic logarithms* (#63), to be executed on a programmable calculator. It was checked out on a Casio fx-7000G, but other calculators may not exhibit the same problems in all parts of the project. This may be a hard project since students are expected to identify and rectify anomalous behaviors without always being given any idea of what these might be.

Pi in the Sky

You are the "super-programmer" for the world's largest supercomputer. Your boss has taken you aside to give you your next assignment. He explains, confidentially, that for several days now, Project SETI—which is concerned with the search for extraterrestrial intelligent life—has been receiving a string of digits from a powerful point source near Tau Ceti. He suspects that these are the digits in the decimal representation of π. However, he doesn't know when they started transmitting, so they may have gotten quite far out, perhaps to the millionth decimal place or beyond. Your assignment is to calculate π to more decimal places than ever before, in order to compare these computed digits with those from the extragalactic transmission. What you need is an algorithm; that is what you will develop here.

a. Express $\tan^{-1} z$ as the definite integral of some function $f(t)$, with z as the upper limit. Show how, if you could obtain a numerical value for this integral when $z = 1$, then you could get an exact numerical value for π.

b. Now write the integrand $f(t)$ as a geometric series. Use the identity for the sum of the first n terms of a geometric series to express the integrand as the sum of the first n terms plus an "error term."

c. Integrate this sum term by term to get an expression for $\tan^{-1} z$ consisting of a polynomial plus an error term. Show that the error term has the form of an integral.

d. Give an upper bound on the error incurred when using only the polynomial of part (c) as your approximation to $\tan^{-1} z$. How many terms do you need to give a value for π accurate to a hundred decimal places? To a million? If the computation of each term and its addition to the previously computed terms takes 1 μs on the supercomputer, then how many years will it take to compute π to a hundred decimal places? To a million? To understand why this is a poor series to compute π, evaluate the first six polynomials at $z = 1$.

e. You mention your problem over lunch to Sylvia, the mathematician down the hall. She jots a couple of formulas down for you:

$$\frac{\pi}{4} = \tan^{-1}\frac{1}{2} + \tan^{-1}\frac{1}{3}, \tag{1}$$

$$\frac{\pi}{4} = 4\tan^{-1}\frac{1}{5} - \tan^{-1}\frac{1}{239}. \tag{2}$$

Verify the correctness of these identities. (It is not sufficient to punch the *arctan* button on your calculator!)

f. How would you convert Sylvia's formulas into algorithms to approximate π? Do her formulas yield better approximations (in terms of time and money) than your original method in parts (a)–(d)? Evaluate the first six polynomials for both parts of (1) and (2) and put these beside the six term approximation in part (d). Compare the number of terms of your original series and the series generated by formulae (1) and (2) needed to approximate π to a million decimal places. How long will these formulae take on the supercomputer to yield approximations to π correct to a million decimal places?

Comments

Requires: infinite series, geometric series

Teaches: infinite series, error estimation, numerical methods, integral inequalities

Students enjoyed this project, especially those who were interested in computing. One student was quite excited that, by knowing a little theory, he might be able to program his personal computer to beat his pocket calculator in accuracy on the trigonometric functions and their inverses.

We rewrote a number of parts in response to students missing the point. In part (a), students often simply wrote $\int \tan^{-1} z \, dz$; now we explicitly mention another function $f(t)$ as the integrand. Adding computations to parts (d) and (f) helped students to appreciate better how slowly or rapidly these series converge.

Tripple Trouble

You are the captain of the Starship Exitprize and tripples have invaded the hold of your ship. You are concerned enough to ask Spook, the Vulgar, for an analysis. As usual, Spook takes a superior attitude. Vulgars are experts with infinite series, although they never developed calculus. Hoping to teach you something, he gives his report in a series of riddles, to which you must supply the answers. Supply answers to questions (a) through (d) below, and then go Spook one better by showing him how Earthlings would use calculus to solve the problem in parts (e) through (h).

Spook speaks

Let y_0 be the population of tripples at some initial time $t = 0$, Captain. Then let y_n be the population after n intervals of length Δt have elapsed, i.e., at $t = n\Delta t$. If you allow one more interval to elapse, the population will have increased during the next interval by

$$\Delta y_n = y_{n+1} - y_n.$$

Suppose Δy_n is proportional to both y_n and Δt:

$$\Delta y_n = k\Delta t\, y_n,$$

where k is the growth rate. (You may neglect the death rate in your calculations.)

a. Give an expression for y_1, the total number of tripples after one interval Δt, in terms of y_0, the initial population at $t = 0$.

b. Give expressions for y_2, y_3, and y_n, the total number after 2, 3, or n intervals respectively, in terms of y_0. Make your expressions as compact as possible and factor out y_0. Rewrite your expression for y_n in terms of the total elapsed time t by setting $\Delta t = \frac{t}{n}$.

c. Now expand the expression for y_n you found in part (b) using the binomial theorem. In order to get better resolution and accuracy, let the time scale of your analysis become finer and finer; that is, let $n \to \infty$ as $\Delta t \to 0$. Assume that you can take this limit term by term; your result is an infinite series. Please write this series.

This series is your answer for the population at time t, Captain. You may calculate the answer to any desired degree of accuracy by including enough terms from the series.

Captain Kork's Comeback

d. How do you know you can calculate the answer to arbitrary accuracy? For example, how far off would I be if I just added up the first hundred terms? The first N

terms? Give me an upper bound for the error after 100 terms, and the error after N terms, Spook, and then prove that the error goes to 0 as N goes to ∞.

Now wipe that superior smile off Spook's face by showing him what can be done with Earthling calculus.

e. Start with the same expression for the increment, $\Delta y = (k\Delta t)y$, as Spook used in part (a), and convert it into a differential equation by taking the limit as $\Delta t \to 0$ and using the definition of the derivative. Solve this equation by separation of variables to find the function $y(t)$, which gives the population at time t in terms of the initial population y_0. It is your turn to smile. (Spook had to run to the ship's library to look up separation of variables.)

Spook's Comeback

Captain, I claim that your solution is no different from mine. You have simply renamed my infinite series as your function because Earthlings cannot handle the concept of infinity. Allow me to demonstrate, Captain. In a book in the ship's library, I came across something you Earthlings call the 'derivative.'

f. Differentiate my series term by term with respect to time and see if the answer isn't k times my series. Therefore, my series does satisfy your differential equation, since its derivative is k times itself. Now compare your function from part (e) with your answer in part (c). Your function and my series must be one and the same!

Kork's Second Comeback

Not necessarily, Spook. The same differential equation may have many different solutions. Allow me to demonstrate ...

At about this time, McCool and Scooter burst into your cabin. Scooter says, "You gentlemen can debate theory 'til you're blue in the face, but that does'na help us with our problem. I'm prepared to use radiation from the ship's engine just one time to exterminate 99% of the little beggars right now. Just give me the word, Captain!"

McCool replies, "I cannot condone such a waste of life, Captain. I've got a drug that will cut their growth rate k in half. I think you'll find this a more effective and less brutal method than the one suggested by our engineer."

g. Decide between Scooter's and McCool's strategies. Which is more effective in the short run? In the long run? Convince them of your answers.

h. In the future, will these two treatments ever result in the same number of tripples again? When?

Comments

Requires: infinite series, binomial theorem

Teaches: infinite series, exponential function, differential equations

Students enjoyed this project. Some extended the story line further and others embellished their report with delightful art work.

Harmonic Summing

An important constant in mathematics is defined by the limit,

$$\lim_{n\to\infty}\left(\frac{1}{1}+\frac{1}{2}+\frac{1}{3}+\cdots+\frac{1}{n}-\log n\right).$$

In this project we are going to show that this limit exists and estimate its value. Set

$$T_n=\frac{1}{1}+\frac{1}{2}+\frac{1}{3}+\cdots+\frac{1}{n}-\log n.$$

It is convenient to work first with a closely related sequence,

$$S_n=\frac{1}{1}+\frac{1}{2}+\frac{1}{3}+\cdots+\frac{1}{n}-\log(n+1).$$

a. Draw a picture using areas to relate the sum of the reciprocals of the first n positive integers to the values of the logarithmic function. Identify $\lim_{n\to\infty}S_n$ geometrically. How can the area giving S_n be enclosed in a sequence of n rectangles, each of length at most one on any side?

b. Obtain an upper bound on the sequence $\{S_n\}_{n=1}^{\infty}$ by summing the areas of the rectangles observed in part (a). Show that the sequence, $S_1, S_2, S_3, \ldots$ is increasing. From this one can conclude that this sequence has a limit, since any bounded increasing sequence has a limit.

c. Prove that if $\lim_{n\to\infty}S_n$ exists, then so does $\lim_{n\to\infty}T_n$.

Hint: Prove that the sequence, $T_n - S_n$, converges to zero. Conclude that the limit in question does exist.

d. You are now going to get a lower bound on the limit of the S_n. On your graph connect with straight line segments the points on the logarithmic curve with integral x-coordinates. The sum of the areas of the resultant triangles is less than $\lim_{n\to\infty}S_n$. Evaluate this sum of triangular areas.

Extra credit: Compute the first few terms in the sequence S_n. You are now going to find a more efficient way to get closer to the limit. Subdivide the intervals of unit length. There will now be new areas to add to your lower bound. Do so, and simplify. This should lead you to an expression with an infinite sum. Evaluate it and thereby obtain a better lower bound.

Subdivide again and repeat this process. Continue to subdivide. Develop a general expression for an arbitrary estimate in the process. These estimates will converge much faster than the original sequence.

A similar process can be started on the other side of the logarithmic curve to reduce the upper bound. Develop this and give a general expression for an arbitrary step. In this way the constant will be sandwiched between two new sequences that come arbitrarily close together. Incidentally you may want to prove for the original sequence that $T_n > T_{n+1}$, and hence that both sequences converge.

Comments

Requires: harmonic series

Teaches: limit of a sequence, estimation

This project has never been assigned. The open-ended nature of the extra credit should encourage students with a creative bent in numerical methods to find spectacular ways of speeding up convergence.

Power Series with Positive and Negative Exponents

A Maclaurin series is a series of the form $\sum_{n=0}^{\infty} a_n x^n$. A familiar example is the geometric series $\sum_{n=0}^{\infty} x^n$, which converges to $\frac{1}{1-x}$ when $-1 < x < 1$. The function $f(x) = \frac{1}{1-x}$ may also be written as a sum of negative powers of x. A little algebra shows that

$$\frac{1}{1-x} = -\frac{1}{x}\left[\frac{1}{1-\frac{1}{x}}\right] = -\frac{1}{x}\sum_{n=0}^{\infty} x^{-n} = -\sum_{n=1}^{\infty} x^{-n}. \tag{$*$}$$

a. For what values of x are the equalities in $(*)$ all valid?

b. Find a Maclaurin series for $f(x) = \frac{1}{x-2}$. What is the region of convergence? Now find a series of negative powers of x for the same function. What is the region of convergence of this series?

Sometimes a series can have both positive and negative powers of x. Such a series,

$$\sum_{n=0}^{\infty} a_n x^n + \sum_{n=1}^{\infty} b_n x^{-n},$$

is said to converge if and only if both $\sum_{n=0}^{\infty} a_n x^n$ and $\sum_{n=1}^{\infty} b_n x^{-n}$ converge.

c. Use partial fractions to write the rational function, $R(x) = 1/(x^2 - 3x + 2)$, as a sum of fractions with linear denominators.

d. Each of the two fractions with linear denominators you found in part (c) will have two power series, one with only nonnegative powers of x and one with only negative powers of x. This will give you four possibilities for a series for $R(x)$. Find those series and their respective regions of convergence.

Comments

Requires: power series, partial fractions, region of convergence

Teaches: power series, region of convergence

This project was assigned in a third semester calculus class. Originally, the name Laurent was used and a number of students copied material from some complex analysis books without understanding what was written.

A Power Series Solution of a Differential Equation

Suppose you are looking for a function f. All you know about f is that it is infinitely differentiable everywhere, always positive, $f(0) = 1$, and $f'(x) = 2xf(x)$ for all x.

a. Use only properties of power series to find a power series representation for $f(x)$. Do not use any other functions you know. Be sure to give a general formula for all the terms. You will need to use mathematical induction to prove your formula is right.

b. Now find a short formula for f in a different way. Pretend you know nothing about power series. Your method and answer should not involve power series in any way.

   Hint: Bring all occurrences of f to one side and integrate.

c. After completing parts (a) and (b), explain why these two representations of f agree.

Comments

Requires: mathematical induction, differentiation of power series, techniques of integration, exponential function, natural logarithm, Taylor series

Teaches: mathematical induction, power series, recurrence formulae, separable differential equations

   Make clear that part (a) is to use power series methods only, and (b) is the opposite. Students may need help in (a) with applying induction and indexing.

Rearranging an Infinite Series

The alternating harmonic series

$$\sum_{n=1}^{\infty} \frac{(-1)^{n+1}}{n}$$

converges to log 2. Now suppose s is any given number. You are going to prove the amazing result that the alternating harmonic series can be rearranged (that is, its terms can be written in a different order) so that the resulting series converges and has s as its sum.

You will need to understand the precise definitions of the limit of a sequence and the sum of a series.

Hint: Start by trying to reorder the terms to alternately overshoot and undershoot s with various new partial sums.

Comments

Requires: infinite series, limit of a sequence

Teaches: conditional convergence, index notation

One of the best features of this project is that at first no one believes it could possibly be true; i.e., it is highly counterintuitive. Proving the result requires considerable facility with index notation, with which students may need help. In fact, students will generally require quite a bit of help with this one.

Cockroach on an Elastic Tightrope

Before beginning the main parts of your project, answer the following question to help you with your project. Suppose you are standing on an elastic rope b meters from the left end and c meters from the right end.

Then suppose the entire rope stretches uniformly, increasing its length by d meters. How far are you now from each end? After answering this, you are ready to begin.

Part A:

Suppose a cockroach starts at one end of a 1000 meter tightrope and runs towards the other end at a speed of one meter per second. At the end of every second, the tightrope stretches uniformly and instantaneously, increasing its length by 1000 meters each time.

1. Does the roach ever reach the other end?

2. If so, about how long does it take?

To answer these questions, proceed as follows. Consider the sequence $\{d_i\}_{i=1}^{\infty}$ where d_i represents the distance the roach still has to go after i seconds, but before the rope does its instantaneous stretch. Find a formula for $\{d_i\}_{i=1}^{\infty}$ and then simplify it. Prove that your formula is valid for all i by using the method of mathematical induction. Read about mathematical induction in a precalculus or calculus book if you are not familiar with it.

Part B:

Suppose you don't know how long the rope is initially, nor how fast the roach runs, nor how much longer the rope gets after every second. All you know is that the roach maintains constant speed, and that the rope stretches uniformly and instantaneously by some fixed amount after every second. Answer the same two questions from Part A.

Extra credit: Suppose that instead of stretching only discretely at the end of each second, the rope is stretching continually at some fixed rate of increase. What happens now? For how long does the roach see the other end of the rope receding? How is your formula for the distance remaining related to the discrete one you found above?

Hint: Taking a limit of the discrete situation as the time interval approaches zero yields a differential equation. Look in a differential equations book under exact differential equations and integrating factors.

Comments

Requires: infinite series

Teaches: infinite series, recurrence formulae, mathematical induction, arbitrary constants

Students (and instructors) liked this project and were quite successful with it. One student found a novel and slick method of solution by studying instead the sequence of successive fractions of the total length covered, which is invariant under stretching.

A Harmonic Series with Missing Digits

If $a_1, a_2, a_3, \ldots$ are the positive integers whose decimal representations do not contain the digit 7, show that

$$\sum_{n=1}^{\infty} \frac{1}{a_n}$$

converges and its sum is less than 90.

Here are some suggestions: break up the sequence of positive integers into carefully chosen blocks (not necessarily all the same size), count the number of terms in each block, and see what fraction of them is left after removing those with an offending 7. Then bound the size of the sum of the reciprocals in each block, and try to compare this result to a convergent series.

Comments

Requires: geometric series, comparison test

Teaches: infinite series, counting arguments

This was originally assigned with just the first sentence. Instructors needed to spend a lot of time helping students. The hints have been added because of this. In fact, the statement "see what fraction ... offending 7" is a big hint, but in light of our experience it seems appropriate.

Series and Products

You are familiar with the use of the sigma notation $\sum_{k=1}^{n} a_k$ to denote the sum

$$a_1 + a_2 + \cdots + a_n.$$

Similarly, we use $\prod_{k=1}^{n} a_k$ for the product

$$a_1 a_2 \cdots a_n.$$

For example,

$$n! = \prod_{k=1}^{n} k.$$

Suppose that the sequence of numbers $\{a_k\}_{k=1}^{\infty}$ satisfies $0 < a_k < 1$ for each k and $\sum_{k=1}^{\infty} a_k = \infty$. Prove that

$$\lim_{n \to \infty} \prod_{k=1}^{n} (1 - a_k) = 0.$$

Hint: Begin by proving that $\log(1 - x) \leq -x$ for $0 \leq x < 1$. Use this and results about logarithms to prove that the limit above is 0.

Comments

Requires: infinite series, natural logarithm

Teaches: infinite products

This project has never been assigned.

With the hint, this is probably a one-week project. We are not sure how students would solve the problem without the hint.

Escape from the Cleangons

The starship Exitprize has been captured by the Cleangons and is being held in orbit by a Cleangon tractor beam. The orbit is elliptical with the planet Cleangon at one focus of the ellipse. Repeated efforts to escape have been futile and have almost exhausted the fuel supplies. Morale is low and food reserves are dwindling.

In searching the ship's log, Mr. Spook discovers that the Exitprize had been captured long ago by a Cleangon tractor beam and had escaped. The key to that escape was to fire the ship's motors at exactly the right position in the orbit. Captain Kork gives the command to feed the required information into the computer to find that position. But, alas, a Cleangon virus has rendered the computer all but useless for this task. Everyone turns to you and asks for your help in solving the problem.

Here is what Mr. Spook has discovered. If F represents the focus of the ellipse and P is the position of the ship on the ellipse, then the vector $\overrightarrow{FP}$ can be written as a sum $\overrightarrow{T} + \overrightarrow{N}$ where $\overrightarrow{T}$ is tangent to the ellipse and $\overrightarrow{N}$ is normal (perpendicular) to the ellipse. The motors must be fired when the ratio

$$\frac{|\overrightarrow{T}|}{|\overrightarrow{N}|}$$

is equal to the eccentricity of the ellipse. Your mission is to save the starship from the Cleangons.

Comments

Requires: ellipse, vectors, eccentricity

Teaches: vector projection, vector algebra, ellipse

Students enjoyed this project and many entered into the spirit of it with their own story line. Most used rectangular coordinates, but one group used polar coordinates with the planet at the pole. It can also be solved using parametric equations for the ellipse. Students should be warned that if the computations get very involved, they have overlooked some means of simplification. In such cases, they should seek help.

Measuring the USO

You are an observer at a SETI (Search for Extraterrestrial Intelligence) station on Planet Mickey, the newly discovered tenth planet in our solar system. Your station is located on the equator of this planet. Planet Mickey is unusual in that its axis of rotation is always perpendicular to the line connecting the center of the planet and the center of the sun.

One morning you observe an unidentified stationary object (USO) in the sky. It is in the shape of a large flat plate and is hovering directly overhead. In response to your report of the sighting of the object, you are requested to determine its size. You can't wait for experts with sophisticated equipment to come from earth to do the measuring, so you decide to find its area using the calculus and vector algebra that you know.

At the instant that the sun is directly above you, the USO casts a shadow on the ground and you can calculate the area of the shadow quite easily. The problem is to use this information to find the surface area of the USO.

a. Begin by looking at the analogue of the problem in the plane. If L is a line segment that is not vertical, find its length by knowing the projection on the x-axis and the angle that L makes with the x-axis.

b. Suppose that you have a planar region R in space that is not perpendicular to the xy-plane and a vector N perpendicular to the plane of R. Let S be the projection of R on the xy-plane. How would you use this information and the result in part (a) to find a formula for the area of R in terms of the area of S? What is the formula? Justify your answers.

By now you realize that just knowing the area of S is not sufficient to find the area of R. However, there are some other possibilities to consider.

c. Suppose that the USO were positioned in such a way that you could find its projections on three mutually perpendicular planes. How would you use the result of (b) to find the area of the USO? Find a simple formula relating all four areas. Where have you seen such a formula before?

Unfortunately, you cannot obtain the three projections, but perhaps a second observation on the ground might be helpful.

d. The USO is oriented so that at some time during the day, the sun's rays are perpendicular to its surface. You have marked on the ground the shadow when the sun is directly overhead. As the planet rotates, the shadow moves. After waiting a bit, you measure the angle of inclination of the sun's rays and mark the outline of the shadow at that time. Use this information to find the area of the USO. You might want to look at the two-dimensional case first. Your answer will involve some measurements. Explain exactly what those measurements are. Explain why your technique works.

Extra credit: In part (d) it was important that the sun's rays were perpendicular to the USO at some time during the day. Explain why the method in part (d) fails if this is not true.

Comments

Requires: vectors, three-dimensional geometry, trigonometry, direction cosines, area

Teaches: vectors, three-dimensional geometry, projections, normal vector

This project is a tough one. Many students found the correct answer to part (b) but couldn't give an explanation. In part (d) they used the words "length of the shadow" without explaining what that meant. You should warn your students against such omissions in advance. A physical model can be quite helpful.

A Mountain Climber

On her way down from Mount Everest, Valerie notices that for every three me
she travels northwest she climbs $\frac{1}{2}$ meter, and that for every two meters she trave ..s
northeast she descends $\frac{1}{4}$ meter. In what direction should she start for fastest descent?
If she travels in this direction at 2 meters per second, what will be her rate of descent?
Derive an expression for her rate of descent as a function of the direction travelled
and the speed in that direction. In what directions should she start in order not to go
up or down?

Comments

Requires: gradient, multivariate chain rule

Teaches: directional derivative

This project was assigned as a two-week project, but may be easy enough to do in
one week.

Studying Desertification

A friend of yours named Maria is studying the causes of the continuing expansion of deserts (a process known as desertification). She is working on a variety of fronts, but of crucial importance is a biological index of vegetation disturbance that she has defined. By seeing how this index and other factors change through time, she hopes to discover the role played in desertification by human activities like livestock grazing, land development, and water use, as well as by both human-caused and natural climatic changes. She is studying a huge tract of land bounded by a rectangle; this piece of land surrounds a major city, but does not include it. She needs to find an economical way to calculate for this piece of land the important vegetation disturbance index $J(x, y)$, which she can determine for every point (x, y) in the rectangle but not in the city. Here x and y represent respectively the east-west and north-south coordinates in the rectangle, whose sides are parallel to the axes, with the city in the center. Unfortunately, determining the actual value of the index J for a particular point is quite time consuming and expensive, requiring a lot of field observation and laboratory work, and she has only actually done this for one location so far.

However, Maria has embarked upon an ingenious approach towards combining the results of photographic and radar images taken during flights over the area to calculate the index J much more easily. She is assuming that J is a smooth function, i.e., it has partial derivatives of all orders. While the flight data does not directly reveal the values of the function J, it gives the rate at which the values of J change as the flights sweep over the landscape surrounding the city. Her staff has conducted numerous east-west and north-south flights, and from this mountain of data she believes she has been able to find actual formulae for the rates at which J changes in these two directions over the entire landscape surrounding (but not including) the city. She has given these functions the names M and N. So $M(x, y)$ is the rate at which J changes as one sweeps in the positive x direction, and $N(x, y)$ is the corresponding rate in the y direction.

Maria proudly shows you the two formulae she has somehow deduced from her flight data:

$$M(x, y) = 3.4e^{x(y-7.8)^2} \quad \text{and} \quad N(x, y) = 22\sin(75 - 2xy).$$

a. Convince her that these two formulae cannot possibly be correct. Do this by showing her that there is a condition that the two functions M and N must satisfy if they are to be the east-west and north-south rates of change of the function J, and that her formulae for M and N do not meet this condition.

Maria is disappointed, but determined not to give up on this method since it seems so promising. After a few weeks she calls you, in a very excited state, and tells you that she has discovered that the problem originated in a flawed computer program that one of her staff had written for interpreting the flight data. Moreover, now that the problem is fixed, she has completely different formulae, and the most exciting thing is that the new formulae she has obtained actually satisfy the condition you showed her was necessary. But by now, Maria has acquired a healthy skepticism. She says to you "Just because my M and N satisfy that condition, why does that ensure that they are really the rates of change for some function J? And how do I find a formula for the function J from the formulae I have for M and N? And how do I know the formula I get for J is the only possible one fitting all the data I have?"

b. Your next task is to answer these three questions affirmatively for her, assuming now that the region under study is a full rectangle (i.e., with no city in the middle excluded from the study). Suppose that the value of $J(x, y)$ is known for one particular point in the rectangle, and that M and N are known for the entire rectangle and satisfy the condition in part (a). Show how J can then be obtained. First show how to find the values of J along either a horizontal or vertical line (your choice) through the one point where it is known: recall how to find a function of one variable if you know its derivative, and its value at some point. Then use the same idea to extend your knowledge of J in the perpendicular direction away from the points on this line. Show that your resulting formula really has the desired properties, i.e., it has the right value at the particular point, and has the correct rates of change everywhere. Remember the condition in part (a); it will be very useful, in fact necessary, here. You may also make use of the theorem from advanced calculus stated at the end of this project. Do you appear to get a different function if you start the above procedure for finding J in the direction perpendicular to the one you first chose? Finally, resolve and explain this by showing that your function J is actually the only possible one by proving that any two possible solutions must in fact be the same.

Hint: Recall that two functions of one variable with the same derivative differ by a constant. Expand this argument to a function of two variables by first holding one variable fixed.

Maria is impressed by all this, but still concerned because your proof seems to rely on the whole rectangle being included in the study (i.e., no excluded city).

c. Explain why your procedure for finding J falls apart if the city is there, i.e., if the city is a region in the middle excluded from the study.

In part (b) above you will want to make use of the following theorem from advanced calculus.

Theorem.   *Suppose $f(x, y)$ is a smooth function on a rectangle in the xy-plane. Then*

$$\frac{d}{dy} \int_a^b f(x, y)\, dx = \int_a^b f_y(x, y)\, dx \quad \text{and} \quad \frac{d}{dx} \int_c^d f(x, y)\, dy = \int_c^d f_x(x, y)\, dy.$$

Extra credit: You happen to mention all this to an advanced mathematics student, and you comment that surely the things you have proven must still be true even if the city is in the middle. The student suggests with a smile that you study the formulae:

$$M(x, y) = \frac{-y}{x^2 + y^2} \quad \text{and} \quad N(x, y) = \frac{x}{x^2 + y^2}.$$

Check the condition of part (a), and try to produce a function J. Use your method from above. Things will be easier if you integrate first along the x-axis from a starting point on the axis. Explain what the results from this example show about Maria's final concern. Warning: recall that a function must have a single value for each point in the domain.

Comments

Requires: partial derivatives, equality of mixed partial derivatives, integration, second fundamental theorem of calculus

Teaches: closed and exact differentials, topology

    A story line is woven around proving that closed forms are exact on a rectangle, and that the proof fails if a point is missing. This was given as a group project.

Tetrahedra

Start with the hyperbola $xy = k$, where k is any arbitrary positive constant. Draw several lines tangent to it at various points in the first quadrant. Calculate the area of each triangle formed by the tangent line and the coordinate axes. What do you notice? Make a conjecture about the areas of these triangles and prove your conjecture using calculus.

Now generalize to three dimensions. Let k be a positive constant and consider the plane tangent to the surface $xyz = k$ at a point in the first octant. Consider the volume of the tetrahedron formed by this plane and the coordinate planes. Based on your investigation of hyperbolae, make a conjecture about how this volume changes as the point of tangency varies? Now prove your conjecture.

Comments

Requires: tangent plane, volume by slices

Teaches: tangent plane

We assigned this project before the tangent plane was formally introduced in class and directed the students to read about the tangent plane on their own. The project was initially assigned without the two-dimensional introduction; since we found that we had to direct almost every student to this as a "learning problem," we decided to include it explicitly. Two other projects, *Equal areas* (#23) and *Invariant areas* (#24), treat the two-dimensional problem more fully.

Angles Between Surfaces

The angle between two surfaces $z = f(x, y)$ and $z = g(x, y)$ at a common point is the angle between the vectors perpendicular to the surfaces at that point. Find a formula for this angle. This procedure can lead to more than one value between 0 and π for the angle between the surfaces at this point. Explain why.

Comments

Requires: tangent plane, normal vector, gradient, dot product

Teaches: vector calculus

This was successfully assigned to students who knew only a little about multivariate differentiation and approximation, and had to learn on their own about tangent planes, normal vectors, and 3-dimensional gradients by reading ahead in their text or other books.

The Intersection Curve of Two Surfaces

Consider a point on the curve of intersection of two surfaces $z = f(x,y)$ and $z = g(x,y)$. Find a formula for a vector tangent to the curve at this point.

Comments

Requires: tangent plane, cross product

Teaches: tangent plane, curves in three-space, arbitrary constants

This is a one-week project. It can be found in many books.

Tangent Planes to Cones

a. Find an equation of the plane tangent to the cone $z^2 = a(x^2 + y^2)$ at the point (x_0, y_0, z_0) on the cone, assuming that this point is not the vertex. Show that every such plane passes through the vertex.

b. More generally, if a surface is given by

$$z = xf\left(\frac{y}{x}\right),$$

show that all its tangent planes have a common point.

c. Show that the result in (a) really is a special case of (b).

d. Give a simple geometric argument for why part (b) is true.

Comments

Requires: tangent plane

Teaches: tangent plane, cones

This project was originally assigned without part (d), which was added to give an intuitive appreciation for what they had proven.

Lumpy Land

You find yourself in a strange undulating landscape given by the function,

$$z = f(x, y) = \cos y - \cos x,$$

where z is the elevation.

a. Find all maxima, minima and saddle points. What are the level curves for $z = 0$? Graph this function, $z = f(x, y)$.

 You are now at the origin, and wish to hike to the point $(4\pi, 0, 0)$. You contemplate two rather different routes.

b. Your first route always keeps you at the same elevation. Determine such a route of minimal length. What is its length?

c. Your second route always moves along the gradient. Determine such a route of minimal length, assuming you start hiking in the positive x-direction. What is its length? If you cannot find an exact answer, then determine an upper bound and a lower bound between which the actual length must lie.

d. Which route is the shorter—that of part (b), or (c)?

 Appropriate pictures should be supplied throughout. Justify your answers.

Comments

Requires: gradient, numerical integration, differential distance formula

Teaches: spatial visualization, arc length

 This project was first assigned without part (a). Students tended to consider the level curves at $z = 0$ as $y = x$ only, and needed to be reminded about the other branches.

 Many students tried Simpson's rule to approximate the integral in part (c), but they had difficulties estimating the error.

Swimming and Swinging

Imagine a swimming pool in the shape of a cube with sides of length 2, in some suitable system of units. Two opposite vertices are at temperature 1°, again in a suitable system of units. The other vertices are all at 0°. A swimmer named Frank is going to swim on the diagonal between the two vertices at the higher temperature. He is invigorated by differences in temperature—he likes cold water; in fact, his acceleration is proportional to the negative of the gradient. We wish to find his position as a function of time. To do this we need a preliminary investigation into Laplace's equation. To make it easier to discuss our strategy let us assume that the center of the cube is at the origin, and the vertices at 1° are located at the points $(1, 1, 1)$ and $(-1, -1, -1)$.

a. Show that the function,

$$T_1(x, y, z) = \frac{1}{8}(x + 1)(y + 1)(z + 1),$$

 is a solution to Laplace's equation, and moreover find its value at all the vertices. This is almost the solution we want.

b. Prove that the sum of any two solutions to Laplace's equation is again a solution. We say that solutions may be *superimposed* when we can do this. (In proving this, do not just use examples!)

c. Find another solution T_2 that is 1° at the opposite vertex as before and zero at all the other vertices.

 Hint: Try reflecting your first solution through the origin so that (x, y, z) goes to $(-x, -y, -z)$. What happens when you add these two solutions together?

d. Find the gradient of the sum, $T = T_1 + T_2$, and evaluate it on the diagonal D along which the swimmer swims. Show that on D, the gradient ∇T is parallel to the diagonal D .

e. Write T as a function of the one variable s that is the signed distance from the origin along the diagonal. Find $\frac{dT}{ds}$ and set this proportional to the negative of the swimmer's acceleration. Solve this differential equation given that Frank starts at $(1, 1, 1)$ at time $t = 0$ with velocity 0, and reaches $(-1, -1, -1)$ half a minute later. In the course of swimming back and forth along the diagonal, how long at most does Frank have to hold his breath?

Comments

Requires: differential distance formula, gradient, Laplace's equation

Teaches: Laplace's equation, differential equations, spatial visualization

When this project was first assigned, students tended to pick examples of two functions that satisfied Laplace's equation and to show that their sum did also, instead of writing a proof to answer part (b). This part has been amended to prevent this.

They also needed to be reminded of the meaning of "proportional" in part (e).

An Ant in the Sugar Bowl

An ant at the bottom of an almost empty sugar bowl eats the last few remaining grains. It is now too bloated to climb at a vertical angle as ants usually can; the steepest it can climb is at an angle to the horizontal with a tangent equal to 1. The sugar bowl is shaped like the paraboloid,

$$z = x^2 + y^2 \qquad (0 \le z \le 4),$$

where the coordinates are in centimeters.

a. Find the path the ant takes to get to the top of the sugar bowl, assuming it climbs as steeply as possible. Use polar coordinates (r, θ) in the xy-plane, and think of the ant's path as parameterized by r; then find a relation between the differentials $d\theta$ and dr, and integrate this relation to get $\theta(r)$.

b. What is the length of the ant's path from the bottom to the rim? To answer this, first discover a formula for arc length involving dz, dr and $d\theta$ in three dimensions.

c. Draw a graph of the sugar bowl and the path the ant takes to get out.

Hint: You may want to start with the projection $\theta(r)$ of the path in the $r\theta$-plane.

Comments

Requires: differential distance formula

Teaches: cylindrical coordinates, function defined piecewise

This is a hard project. Students had difficulty at first seeing qualitatively how the ant will move. Also, they are unaccustomed to manipulating differentials via the Pythagorean theorem—a useful tool which many modern textbooks have ignored.

Perpendiculars to an Elliptical Paraboloid

Consider all lines that are perpendicular to the surface $z = ax^2 + by^2$ at a fixed height h above the xy-plane. The set of all such lines intersects the xy-plane in a curve. Find an equation for this curve. What familiar type of curve is it?

Extra credit: As h increases without bound, what shape does the curve approach? As h approaches zero, what shape does the curve approach? Is the curve ever a circle? If so, where?

Comments

Requires: tangent plane, perpendicular to a surface

Teaches: curves, lines in space, normal vector, arbitrary constants

This is a one-week project and can be found in some books.

A Flying Runner

Robin is running around on the inside of a bowl-like surface, $z = x^2 + y^2$, with distance measured in meters. It is noon, the sun is directly overhead, and the birds are silent. All of a sudden, much to her horror, she realizes someone has chopped off the bowl at a height of 3 meters. But she is too close to the edge and running too fast, so she flies up off the edge of the bowl at a speed of 5 meters per second. As she flies up off the edge, she looks down at the ground below (the xy-plane) and sees her shadow. She observes that it is in the first quadrant, has x-coordinate 1, and it is heading directly towards a house with coordinates $(2, 4)$. Realizing she has no parachute, she wonders (of course) how long it will be before she hits the ground and where she will land. Set her mind at ease by answering these two questions for her. Ignore air resistance.

Hint: Read about things like simple motion under the force of gravity, parametric equations, and velocity vectors.

Comments

Requires: vectors, parametric equations, motion under gravity

Teaches: vector calculus

Students will need to know or read about things in the hint. Students without vector calculus will need to learn how to combine 3-dimensional vectors with motion under gravity. The project was assigned without the hint. We believe the hint is necessary to cut down on instructor time helping students.

Seeking the Greatest Increase in Temperature

You are standing bare-footed on the xy-plane at the point $(-2, 1)$. The temperature of the plane at each point (x, y) is given in degrees Celsius by

$$T(x, y) = 10 + x^2 - y^2.$$

You are going to move around on the plane seeking a warmer location. Your strategy is very simple: always move in the direction of fastest temperature increase.

a. First, find the curve you will move along. You will need to read about separable differential equations.

b. Suppose you are confined to the disc of radius 5 centered at the origin. Where will you end up, and will it be a comfortable place to stand?

c. Suppose you are no longer confined to this disc. Where will you end up? How hot will your feet be then? Will life become unbearable, and if so, where?

Comments

Requires: gradient, techniques of integration

Teaches: gradient, separable differential equations

The original project, as used in class, was essentially parts (a) and (c).

Minimal Pentagon

Consider the cross section of a house:

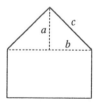

We want to minimize the perimeter subject to the area being 30 m^2. This is an attempt to minimize the cost of a house of given size.

a. Solve this minimization problem by the method of Lagrange multipliers. Use the Pythagorean theorem,

$$a^2 + b^2 = c^2, \qquad\qquad (*)$$

to eliminate one of the variables, thus reducing the problem to three variables with one constraint and one multiplier.

b. In part (a), the calculations were difficult because of the expressions involving radicals. To avoid these difficulties, solve this problem again by leaving $(*)$ as a second constraint. Begin by extending the theory of Lagrange multipliers to allow for a second constraint and a second multiplier. Then solve the specific problem above using this idea.

Comments

Requires: Lagrange multipliers

Teaches: Lagrange multipliers

This problem was assigned as a two-week project in a course for biologists and social scientists. This would probably be only a one-week project in a calculus course for engineers, physical scientists, and mathematicians. However, there are three-dimensional analogs that are more challenging.

Some textbooks only discuss the case of one multiplier; so, given the typical textbook in such courses, students have to figure out how to pass from one multiplier to two, but this did not cause our students much difficulty. In theory, par. harder since it involves more equations, but in practice it is just as easy.

An Unusual Critical Point

A point (x_0, y_0) is called a *saddle point* for a function f if (x_0, y_0) is a critical point for f, but f has neither a local minimum nor a local maximum at (x_0, y_0).

Consider the function

$$f(x, y) = (y - x^2)(y - 2x^2).$$

a. Establish that $(0, 0)$ is a critical point for f. Does f have any other critical points?

b. Show that the second-order partial derivative test fails to establish whether f has a local maximum, local minimum, or a saddle point at this critical point.

c. Examine the behavior of $f(x, y)$ along every line in the xy-plane passing through the origin. Does anything special happen there?

d. What does f have at the origin, a maximum, a minimum, or a saddle point?

   Hint: Where is f positive, where is f negative, and where is f equal to 0?

e. Explain intuitively what is going on. Explain what is strange here and how to reconcile it. Graph f in three dimensions, and explain in words what the graph looks like.

Comments

Requires: Hessian

Teaches: spatial visualization, saddle point

The beauty of this project is its counterintuitive appeal. Once this is seen, it is easy to construct similar examples: $f(x, y) = (y - x^3)(y - 2x^3)$, or the more exotic $f(x, y) = (y^2 - x^3)(y^2 - 2x^3)$.

This project was assigned in two courses with rather different clientele. The first consisted of students in engineering and physical science, who needed only a modest amount of help. The second class consisted of students in biology and the social sciences, who found it very difficult; the idea of examining where the function is zero was totally foreign to them. Only one of these students was able to graph the surface, and then only with a three-dimensional plotting routine on a computer.

Peculiar Critical Points

There are surfaces with saddle points that would be hard to put on a horse and impossible to ride. This project looks at two such surfaces.

A point (x_0, y_0) is called a *saddle point* for a function f if (x_0, y_0) is a critical point for f but f has neither a local minimum nor a local maximum at (x_0, y_0).

I. Consider the function

$$f(x, y) = x^3 - 3xy^2.$$

a. Establish that $(0, 0)$ is a critical point for f. Does f have any other critical points?

b. Show that the second-order partial derivative test fails to establish whether f has a local maximum, local minimum, or a saddle point at this critical point.

c. Examine the behavior of $f(x, y)$ along every line in the xy-plane passing through the origin. Does anything special happen there?

d. What does f have at the origin: a maximum, a minimum, or a saddle point?

   Hint: When is f positive, when is f negative, and where is f equal to 0?

e. Explain intuitively what is going on. Explain what is strange here and how it is different from what you have seen previously. Graph f and explain in words what the graph looks like.

II. Having discussed that peculiar function, you should have no trouble with the new function

$$g(x, y) = xy^2.$$

Repeat all the steps above for g. Are the graphs of both functions similar? If so, how?

Comments

Requires: Hessian

Teaches: spatial visualization, saddle point

Students in biology and social sciences, to whom this project was assigned, found it difficult and frustrating. Zero sets and three-dimensional graphing are definitely not their cup of tea. Probably one of these functions alone is enough for them.

Tracking Missiles

Suppose a missile is launched at White Sands Missile Range and the data measured for its course are as follows.

t (time)	x	y	z (height)
0	1	0	3
1	48	32	116
2	103	57	193
3	145	91	237

Time is measured in seconds and distance in feet.

The object of this project is to adapt the method of least squares to predict the future course of this missile. We want equations describing its motion. In particular, what will be its maximum height, where will it land, and when? You may assume that the only force acting on it is gravity.

a. Decide what kind of equations will be needed to describe the orbit. These will have some parameters yet to be determined numerically. What is important now is the form of the equations. To get started, read about falling bodies and motion in the plane.

b. Are these equations directly amenable to using the method of least squares to determine their parameters? If not, what modifications have to be made?

c. Find the parameters and the equations describing the path.

d. Answer all the questions in the second paragraph.

e. In three dimensions, graph the points given and your approximation to the path.

Comments

Requires: method of least squares

Teaches: parametric equations, spatial visualization

This project was assigned in a course for students in biology and social sciences. Because of their weak or nonexistent backgrounds in physics, many had trouble even getting started with part (a). They had difficulty sorting out dependence among the variables.

Surprisingly, there are many ways to solve this problem and get approximately equal answers, depending on where and how one uses the method of least squares.

Planning a Community

You are a city planner designing a planned community. You are trying to decide between a circle and various regular polygons for the shape of the community. The shopping mall is in the center. You want to calculate the average distance from a resident to the mall, and how this varies with the number of sides. Assume that the population density is uniform.

a. Find the average distance to the mall from a resident in a circular community of radius 1 kilometer.

b. Find the average distance from a resident to the center in a regular polygonal community with n sides, where the distance from the center to the midpoint of each side is 1 kilometer.

c. Explain how the result of part (a) can be derived from that of part (b).

d. Which design gives the most efficient community?

Comments

Requires: average value of a function, l'Hôpital's rule, techniques of integration

Teaches: limits, geometry

The project statement does not specify whether the students need to make the geometrical limit argument precise or rigorous; this is at the discretion of the instructor. Students are explicitly led into a similar formal argument in the project *Finding the area of a circle* (#17).

This project was originally assigned as a pure mathematics problem without the story line.

$$\sum_{n=1}^{\infty} \frac{1}{n^2} = \frac{\pi^2}{6}$$

Here is a proof of this amazing fact.

Consider the integral

$$\int_0^1 \int_0^1 \frac{dx\, dy}{1 - xy}.$$

In part (a), you will show that it equals $\sum_{n=1}^{\infty} \frac{1}{n^2}$. Then in part (b), you will evaluate the integral and find that it is exactly $\frac{\pi^2}{6}$. In part (c) you may become a celebrity.

a. Review the theorem for the sum of a geometric series. Also recall that you know how to integrate a power series. Use these to show that

$$\int_0^1 \int_0^1 \frac{dx\, dy}{1 - xy} = \sum_{n=1}^{\infty} \frac{1}{n^2}.$$

b. The goal now is to evaluate

$$\int_0^1 \int_0^1 \frac{dx\, dy}{1 - xy}.$$

and get $\frac{\pi^2}{6}$. To do this, first make a two-dimensional change of variables by writing x and y in terms of u and v, where the positive u and v axes are obtained by rotating the positive x and y axes respectively by $\frac{\pi}{4}$ in the counterclockwise direction (draw a picture; you'll need it).

Be sure to read about the Jacobian to see how to make a valid change of variables in a double integral. You will find that the nature of the region of integration causes your new integral to break up into pieces. Now use what you know about integration techniques to compute these integrals. As a final hint, if while evaluating an integral you come across $\cos\theta/(1 + \sin\theta)$, you might find it useful to know that

$$\frac{\cos\theta}{1 + \sin\theta} = \frac{\sin\left(\frac{\pi}{2} - \theta\right)}{1 + \cos\left(\frac{\pi}{2} - \theta\right)} = \frac{2\sin\left\{\frac{1}{2}\left(\frac{\pi}{2} - \theta\right)\right\}\cos\left\{\frac{1}{2}\left(\frac{\pi}{2} - \theta\right)\right\}}{2\cos^2\left\{\frac{1}{2}\left(\frac{\pi}{2} - \theta\right)\right\}}.$$

c. Show that

$$\int_0^1 \int_0^1 \int_0^1 \frac{dx\, dy\, dz}{1 - xyz} = \sum_{n=1}^{\infty} \frac{1}{n^3}.$$

So if you could evaluate the triple integral, you would know the sum of $\sum_{n=1}^{\infty} \frac{1}{n^3}$. Its value has remained a great unsolved question ever since Leonhard Euler first posed it in the year 1736. You will be a celebrity if you find it!

Extra credit: Does this technique generalize to $\sum_{n=1}^{\infty} \frac{1}{n^k}$, for k any positive integer?

Comments

Requires: Jacobian, integration of power series, techniques of integration

Teaches: multiple integrals, Jacobian

This was given as a two-week group project. The students felt a real sense of accomplishment upon completing the project.

The Sphere in 4-Space

Your task is to find the volume of the sphere in 4-space. Begin modestly by using a double integral to find the area of the circle in 2-space. Notice how trigonometric substitution is used to evaluate the integral. Next use a triple integral to find the volume of the sphere in 3-space. Again, observe how the integration proceeds.

Now, consider rectangular coordinates (w, x, y, z) in four-dimensional space. Set up a quadruple integral for the four-dimensional volume of the solid ball inside the sphere $w^2 + x^2 + y^2 + z^2 = a^2$ of radius a in 4-space. Then evaluate it.

Extra credit: Derive formulae for the volume of the sphere of radius a in n-space.

Comments

Requires: multiple integrals, trigonometric substitution

Teaches: geometry of n-space

This is a one-week project. If one assigns the extra credit part, it could be a two-week project. For the extra credit part you may wish to tell your students that the formula for n odd and the formula for n even are different.

A Leaking Can

A cylindrical can with constant density is filled with your favorite drink and is sitting upright on the xy-plane. Unfortunately, it has a slow leak in the bottom.

a. Where is the center of mass of the can plus drink before the drink starts to trickle out? What do you think will happen to the center of mass as the drink flows out? Will the center of mass continue to sink as the drink empties? Where is the center of mass when the drink is all gone? Use intuitive arguments to support your answers to these questions.

Now we will use calculus to answer these questions.

b. Suppose R is an object in space divided into two pieces, R_1 and R_2. R does not necessarily have uniform density. Let M_1 and M_2 denote the masses of R_1 and R_2 respectively. Suppose $(\bar{x}_1, \bar{y}_1, \bar{z}_1)$ is the center of mass of R_1 and $(\bar{x}_2, \bar{y}_2, \bar{z}_2)$ is the center of mass of R_2. Find formulae for the coordinates $(\bar{x}, \bar{y}, \bar{z})$ of the center of mass of R in terms of $\bar{x}_1, \bar{x}_2, \bar{y}_1, \bar{y}_2, \bar{z}_1, \bar{z}_2, M_1$, and M_2. Show that $(\bar{x}, \bar{y}, \bar{z})$ lies on the line between the other two centers of mass.

c. Use the result of part (b) to give a formula for the height of the center of mass of the can plus drink when your remaining drink is at height z. Find out how high the drink is when the center of mass of the can plus remaining drink is lowest. Show that at that moment, the center of mass is at the top of the drink. Explain what is going on.

Extra credit: Can you develop an argument not using calculus that shows that the center of mass must reach a minimum, and that the minimum coincides with the top of the drink?

Comments

Requires: center of mass

Teaches: minimization, arbitrary constants

This project led students to compare physical intuition about moments and centers of mass with the results of calculation. Part (b) is an easy proof starting from the definition of the center of mass as an integral. Some books may first define the center of mass as the point on a body at which all the mass may be assumed to be concentrated for the purpose of computing first moments. In this case, there is nothing to do in part (b). To salvage the exercise, first define the center of mass as a quotient of two integrals.

Regions for a Repeated Integral

Devise a region R in the plane such that $\int_R e^{y^3} \, dA$ can be evaluated with the aid of a repeated integral. What is the most general region R you can find for which this is possible?

Comments

Requires: multiple integrals

Teaches: multiple integrals

This project has an open-ended nature in which students are asked to think about various possible answers. Here it is perhaps more important for them to think about the idea than to come up with the most general answer. Instructors will have to decide what is expected for grading purposes. This project is easy enough that it is probably best as a one-week project.

Density in Cubes

a. Suppose that the density of a cube is proportional to the square of the distance from one corner. Show that the mass is the same as it would be if the density were constant and equal to the original density at another corner adjacent to the first.

b. Suppose that the density of a cube is proportional to the square of the distance from the center of the cube. Show that the mass is the same as it would be if the density were constant and equal to the original density at the middle of an edge.

c. Use part (a) to prove part (b) without calculating another integral.

Comments

Requires: multiple integrals, mass

Teaches: mass

Part (a) was assigned as a two-week project, but was too easy.

Rolling Stones

Three cavemen have carved three rocks into three shapes: a sphere, a disc, and a hoop, i.e., a disc with a circular hole in the middle. Assume a reasonable degree of workmanship: these carved shapes are round enough that they roll easily, but still rough enough that they do not slip on the ground. Also assume that each rock has the same radius and the same density throughout. The cavemen start playing a game of letting them roll down a curved hill to see which gets to the bottom first. But it's not much of a contest. One stone always gets there first, and another always last. Assuming Newtonian mechanics, explain with a complete mathematical argument why this is so.

Suppose instead that the three stones are sent uphill, all with the same initial velocity. Which goes the farthest? And which gets to its maximum height the first?

Extra credit: To make a fair contest, let the radii be different. What should their relative sizes be to insure that all three shapes get to the bottom of the hill at the same time? With these new radii what will happen in the second paragraph?

Comments

Requires: moment of inertia, cylindrical coordinates, spherical coordinates, physics

Teaches: moment of inertia, arbitrary constants

This project has never been assigned. It requires surprisingly little calculation once the moments of inertia are found, providing one is willing to do a thought experiment. (N.B. Moments of inertia of common bodies like these are available in tables.) Some knowledge of elementary physics is needed, e.g., the conservation of total energy. The student who plunges in without thinking ahead will be at a serious disadvantage. For example, one must figure out how to handle a hill whose slope is not constant.

Averaging an Affine Function

A function f defined on $\mathbb{R}^3$ is called an *affine* function if it has the form

$$f(x, y, z) = ax + by + cz + d$$

where a, b, c, and d are constants. Suppose S is any region in space. Show that the average value of an affine function on S is just the value of the function at the centroid of S.

State the converse of this result for functions of one variable that have continuous second derivatives everywhere. Prove that the converse is true for any function whose second derivative is identically zero. Then prove that the hypothesis of the converse is not satisfied by any function whose second derivative is not zero at some point. Is all this sufficient to prove the converse?

Comments

Requires: centroid, multiple integrals, Taylor's theorem

Teaches: Taylor's theorem, centroid

This project was originally assigned without the paragraph concerning the converse. It is probably a one-week project.

Jacobi's Pizzeria

In order to stay in school and take more calculus, you have to take a summer job. You have been hired by Jacobi's Pizzeria as a consulting engineer to help them test and calibrate a new automated olive pitter, which they are considering buying. The advertising flyer for the Pits Co. says:

> Stage one of our auto-pitter has two cutting blades spaced 6 mm apart. Olives of various shapes and sizes (but always greater than or equal to 6 mm in every direction) are fed between the horizontal blades, cutting off congruent caps. In stage two a cylinder is punched out of each olive by a combination of pressure and vacuum, taking the pit away. The bottom and top of this cylinder are the ellipses left where the symmetrically placed caps were first lopped off. The auto-pitter has been designed to yield a pitted olive of constant volume independent of the diameter of the olive.

Now Mama Jacobi is very skeptical of this last statement. Before she even orders the auto-pitter, she has hired you to model mathematically its effect on olives of various shapes. In order to use calculus, you assume the olives are all ellipsoids. You go to Papa Jacobi (who in his youth did some work of his own in calculus!), who shows you a method involving coordinate transformations. He then shows you how to apply his method of change of variables in a double integral to make part (a) below even easier by transforming it to polar coordinates in the xy-plane.

Read about the Jacobian and use Papa Jacobi's method for double integrals on parts (a) through (c). For example, for part (b) transform the integral to polar coordinates; for part (c), in order to be able to use the technique of part (b), transform the elliptical cross section to a circle.

a. Your first set of model olives are perfect spheres of diameter at least 6 mm. Show that, surprisingly enough, the company's claim about volume is accurate! Find the volume of the pitted olives.

b. Your next set of model olives are spheroids—they have an elliptical vertical cross section of at least 6 mm in length and a circular horizontal cross section. They are fed between the blades, giving off caps with circular edges. Is the company's claim still valid?

c. Next, you feed your machine an ellipsoidal olive, the skin of which has the equation

$$\frac{x^2}{a^2} + \frac{y^2}{b^2} + \frac{z^2}{c^2} = 1.$$

Find the volume of the pitted olives. Is it constant, i.e., independent of the parameters a, b, and c?

You are now preparing a report of your results to Mama Jacobi. You find that you must do the following.

d. Explain Papa Jacobi's method. Find an appropriate theorem on changing variables in a double integral by means of the Jacobian, and explain how this theorem applies to the changes of variables in the previous parts of this project. Does the theorem you cited cover the entire region of integration?

e. Prove the theorem of part (d). Let the equations,

$$\begin{cases} x = x(u, v) \\ y = y(u, v), \end{cases}$$

be an arbitrary change of variables in a double integral. Use the particular changes of variables you employed in the earlier parts as particular examples to illustrate each step in your explanation of the proof of Papa Jacobi's method.

Mama and Papa Jacobi now offer you a permanent job, promising you won't have to do calculus anymore. Do you take it, or go back to school?

Comments

Requires: Jacobian, multiple integrals

Teaches: Jacobian

One student reported a sudden insight while gorging himself at the local pizzeria.

Students may do this project in ways that do not require the Jacobian, but in part (d) they must come to grips with it in any case.

Variations of this project can be created. One might ask for the integrals to be done by slices perpendicular to the z-axis—suggested by contemplating real slices of olives on real pizza. More generally, it suffices to assume that the three axes of the ellipsoid are in constant ratios, $a : b : c$, for the pitted volume to be constant. That is, if two sufficiently large ellipsoids are similar, and passed through this pitting machine with the same orientation, then the pitted volumes are equal. Thus new but related projects might be created. For example, a harder project would be to find out how the blade spacing should be varied to accommodate different size olives so that the pitted volume is constant.

Transforming College Football

You have been hired as a consulting engineer on the project to design a new 'Rah-Rah Dome' stadium for your record-setting college football team, which has just been endowed with several million dollars after an unprecedented winning streak of two games. The regents have decreed that the stadium is to be completely enclosed, and furthermore that it is to be shaped like half a football. In fact, the equation of the roof of the dome is to have the general form,

$$f(x,y) = 25 \left(1 - \frac{x^2}{a^2} - \frac{y^2}{b^2} \right)^{3/2},$$

measured in meters, with f being 0 at ground level.

The constants a and b will be determined at the end of the calculation, but first . . . .

a. Set up a double integral that gives the volume of the stadium in terms of the constants a and b. Be sure to specify the region R of integration in the xy-plane.

b. Your next task is to perform a change of variables,

$$\begin{cases} x = x(s,t) \\ y = y(s,t), \end{cases}$$

that transforms the region R in the xy-plane to the unit disk (i.e., the unit circle with its interior) in the st-plane. Read about the Jacobian and transform your integral appropriately.

c. Next, transform the resulting integral again, to an integral in polar coordinates (r, θ) in the $r\theta$-plane. Evaluate the integral to get an answer in terms of a and b.

d. Justify your transformations in parts (b) and (c) by stating the theorem on the change of variables in a double integral via the Jacobian, and explain how this theorem applies to them.

e. Find the smallest values of a and b given that the stadium is to be twice as long as it is wide and that it is to hold 20,000 screaming fans. Assume each fan occupies one quarter of a square meter, and the area taken up by the field, including end zones and sidelines, is 10,000 m^2. What is the stadium's volume?

You now take your calculations to your supervisor, who turns out to be none other than your calculus teacher! (Is this a bad dream or what?) He asks you the following questions.

f. The theorem you cited in part (d) probably does not apply to all of R. Where does it fail? Why? How can you argue around this to still use the theorem to justify your work?

g. By the way, will a football playing field really fit into your stadium? Is the stadium really shaped like half a football?

h. Is there a rectangular field of 10,000 m² that fits into your stadium?

Extra credit: Sketch an argument for the truth of the theorem found in part (e). Illustrate each step in your argument for a general mapping,

$$\begin{cases} x = x(u, v) \\ y = y(u, v). \end{cases}$$

Use the specific mappings of parts (b) and (c) as examples.

Comments

Requires: Jacobian, multiple integrals

Teaches: Jacobian

This project was inspired by the New Mexico State University football team losing 18 games straight. One student wrote on her project that she hoped the new dome wouldn't be built, but rather the money spent on mathematics professors.

Students may do this project in ways that do not require the Jacobian, but in part (d) they must come to grips with it in any case. Alternatively, as another project, this could be done with slices perpendicular to the z-axis providing we first do a change of variables to get rid of the $\frac{3}{2}$ power.

This project assumes students know the formula for the area of an ellipse. It might make more sense to start the project off by having them first reduce, via a couple of Jacobians, the problem of finding the area of an ellipse to that of finding the area of a circle in polar coordinates.

For the last part of part (h), students should notice that the lack of intersection between the hyperbola $xy = 10,000$ and the ellipse that forms the base of their dome means that a rectangular field will not fit into their ellipse. Thus the 10,000 m² field in part (e) must have rounded corners to fit.

For the record, the official size of a football field is 120 yards by 160 feet.

The Logistic Equation

Differential equations are important in many sciences. They arise as continuous models, sometimes of discrete processes. There are many techniques for studying them: exact solutions, stability analysis, and numerical approximation. By investigating a class of closely related equations in population ecology, you will travel over a sinuous path, typical of mathematical biology, to gain some understanding of the dynamics of change.

I. A Discrete Model

a. Consider a population, initially of size N_0, that reproduces at the end of each year, increasing by a factor of r. Write an equation that describes the growth of this population $N(t)$ as a function of time t, a nonnegative integer.

b. Let $r = 0.2$ and $N_0 = 100$. Complete the following table and plot $N(t)$ versus t.

t (years)	$N(t)$
0	100
1	.
2	.
3	.
4	.
5	.

c. Consider a different population, initially of the same size 100, whose annual rate r of population increase is now divided over the four quarters of the year so that the population reproduces at the end of each quarter (3 months). Write an equation that describes the size of this population as a function of time t, in years. Plot $P(t)$ versus t on the same graph that relates $N(t)$ to t.

d. Consider yet another population S, initially of size 100, that increases at the same rate r divided over n intervals; that is, it reproduces at the end of n intervals per year. Write an equation that describes the size of this population at the end of each year.

II. A Continuous Model

An important step in mathematical modeling is to pass from the discrete to the continuous. This we do next.

a. Write a continuous function for population growth based on the discrete equation given in part (I.d). Graph this function on the same graph as $N(t)$ and $P(t)$.
   Hint: Let n go to ∞.

b. Continuous population growth described in part (II.a) can be modelled by the following differential equation:

$$\frac{dN}{dt} = rN, \tag{1}$$

where N is the number of individuals in the population. Solve this differential equation by separating its variables, N and t, and integrating each side. Show that this solution gives the same population growth equation as in part (II.a).

III. Limited Resources

The above differential equation describes the growth of a population when resources, such as food, are unlimited. This is rarely the case in nature—most populations have limited resources.

a. Modify equation (1) so that the growth rate r decreases linearly as a function of N, that is, r in equation (1) of part (II.b) is replaced by the expression $r_0 - \alpha N$. Call this new growth rate $r(N)$.

b. Argue qualitatively that the environment cannot sustain a large population of more than K individuals over the long term, where K is some constant.
   Hint: If N is very large is the population growing or shrinking? The constant K is often referred to as the *carrying capacity* of the environment. Relate K to the derivative of the function r of N found in part (III.a). Use this relationship to introduce K into the differential equation of part (III.a).

c. Draw a vertical axis for the population N, including the values 0 and K. On this axis draw arrows going in the direction of population growth or decrease—growth being an arrow upward, and decrease downward. Can you argue on the basis of your picture how the size of the population behaves when it starts at a value slightly above 0? Do you expect the population ever to exceed K? What would happen if the population were started above K?

d. Solve the above differential equation by separating variables and find N as a function of t.

e. Let $r_0 = 0.2$, $N_0 = 100$, and $K = 150$. Plot $N(t)$ versus t. Replot the old $N(t)$ of part (II.a) on this new graph and compare it with the new $N(t)$.

f. Compare your analytical results in parts (III.d) and (III.e) with your qualitative predictions in part (III.c). Were your hunches born out by the analysis?

IV. Stability

a. Assume that $N_0 < K$. With the solution found in part (III.d), prove that $N(t)$ goes to K as t goes to ∞.
b. Assume that $N_0 > K$. With the solution found in part (III.d), prove that $N(t)$ goes to K as t goes to ∞. For $N_0 = 200$ and $K = 150$, plot $N(t)$ on the graph of part (III.e).

Comments

Requires: separable differential equations

Teaches: separable differential equations, discrete and continuous models

This was assigned as a group project in a second course of calculus for a good group of students in the biological and social sciences, simultaneously with lectures on differential equations. They still needed to be helped through certain key steps. With hindsight, it would have been easier for the students to have had it assigned afterwards. It will help students to be exposed to compound interest. The project has been modified somewhat in light of our experience.

The Logistic Equation with Delay

Differential equations are important in many sciences. They arise as continuous models, sometimes of discrete processes. There are many techniques for studying them: exact solutions, stability analysis, and numerical approximation. By investigating a class of closely related equations in population ecology, you will travel over a sinuous path, typical of mathematical biology, to gain some understanding of the dynamics of change.

Read about the logistic equation and study its solution. If we introduce a delay in the time response, then the growth rate of the population becomes:

$$\frac{dN}{dt} = rN\left[1 - \frac{N(t-T)}{K}\right]. \tag{1}$$

Here $N(t)$ is the population as a function of time, r is the growth rate parameter and K denotes the carrying capacity. This differential equation has no exact solution. It will be necessary to use numerical methods, thus returning to the discrete!

a. By rescaling the time t (and changing N to x) show that the equation above is equivalent to a simpler equation. To do this, introduce a new dimensionless variable τ that is r times your old time variable, $\tau = rt$. Let a new dimensionless variable x measure population as a proportion of carrying capacity, $x = \frac{N}{K}$. Transform equation (1) into a new equation in terms of the variables x and τ. You should obtain:

$$\frac{dx}{dt} = x(\tau)[1 - x(\tau - \alpha)]. \tag{2}$$

What is α in terms of r, K, and T? You will find this to be the crucial parameter. Assume that r, K, and T are always positive.

b. For each of the six values below, use the computer program supplied to the class to plot $x(\tau)$ as a function of τ :

$$\alpha = 0.1, 0.2, 0.5, 1.0, 2.0, 4.0.$$

c. Describe what you see.

d. Find the dividing lines between essentially different qualitative behaviors. Try to express these crucial values of α in terms of known constants such as e and π.

Comments

Requires: differential equations, computer

Teaches: differential equations

   This was assigned as an extra credit problem after the preceding project, *The logistic equation* (#102). A program, together with instructions for its use, was written for students to see the graphical solutions of (2). Unfortunately the software would not transfer from a faculty office to the mathematics computer laboratory because of incompatible graphics cards, even though the computers were different models made by the same manufacturer. Since the class was quite small, the students were invited to use—by appointment—the personal computer in a faculty member's office!

   Part (d) is harder for students than it might appear. They weren't sure what qualitative behaviors to look for. For example, they had trouble distinguishing between slow relaxation to a stable equilibrium and the unfolding of oscillatory instability. As instructors we may have forgotten how carefully trained we were to spot such phenomena.

Index